INTERNATIONAL WILDLIFE ENCYCLOPEDIA

THIRD EDITION

Volume 21

WAT–ZOR

Marshall Cavendish Corporation
99 White Plains Road
Tarrytown, New York 10591–9001

Website: www.marshallcavendish.com

Library of Congress Cataloging-in-Publication Data

Burton, Maurice, 1898-
International wildlife encyclopedia / [Maurice Burton, Robert Burton] .-- 3rd ed.
p. cm.
Includes bibliographical references (p.).
Contents: v. 1. Aardvark - barnacle goose -- v. 2. Barn owl - brow-antlered deer -- v. 3. Brown bear - cheetah -- v. 4. Chickaree - crabs -- v. 5. Crab spider - ducks and geese -- v. 6. Dugong - flounder -- v. 7. Flowerpecker - golden mole -- v. 8. Golden oriole - hartebeest -- v. 9. Harvesting ant - jackal -- v. 10. Jackdaw - lemur -- v. 11. Leopard - marten -- v. 12. Martial eagle - needlefish -- v. 13. Newt - paradise fish -- v. 14. Paradoxical frog - poorwill -- v. 15. Porbeagle - rice rat -- v. 16. Rifleman - sea slug -- v. 17. Sea snake - sole -- v. 18. Solenodon - swan -- v. 19. Sweetfish - tree snake -- v. 20. Tree squirrel - water spider -- v. 21. Water vole - zorille -- v. 22. Index volume.
ISBN 0-7614-7266-5 (set) -- ISBN 0-7614-7267-3 (v. 1) -- ISBN 0-7614-7268-1 (v. 2) -- ISBN 0-7614-7269-X (v. 3) -- ISBN 0-7614-7270-3 (v. 4) -- ISBN 0-7614-7271-1 (v. 5) -- ISBN 0-7614-7272-X (v. 6) -- ISBN 0-7614-7273-8 (v. 7) -- ISBN 0-7614-7274-6 (v. 8) -- ISBN 0-7614-7275-4 (v. 9) -- ISBN 0-7614-7276-2 (v. 10) -- ISBN 0-7614-7277-0 (v. 11) -- ISBN 0-7614-7278-9 (v. 12) -- ISBN 0-7614-7279-7 (v. 13) -- ISBN 0-7614-7280-0 (v. 14) -- ISBN 0-7614-7281-9 (v. 15) -- ISBN 0-7614-7282-7 (v. 16) -- ISBN 0-7614-7283-5 (v. 17) -- ISBN 0-7614-7284-3 (v. 18) -- ISBN 0-7614-7285-1 (v. 19) -- ISBN 0-7614-7286-X (v. 20) -- ISBN 0-7614-7287-8 (v. 21) -- ISBN 0-7614-7288-6 (v. 22)
1. Zoology -- Dictionaries. I. Burton, Robert, 1941- . II. Title.

QL9 .B796 2002
590'.3--dc21

2001017458

Printed in Malaysia
Bound in the United States of America

07 06 05 04 03 02 01 8 7 6 5 4 3 2 1

Brown Partworks
Project editor: Ben Hoare
Associate editors: Lesley Campbell-Wright, Rob Dimery, Robert Houston, Jane Lanigan, Sally McFall, Chris Marshall, Paul Thompson, Matthew D. S. Turner
Managing editor: Tim Cooke
Designer: Paul Griffin
Maps: Dax Fullbrook, Seth Grimbly
Picture researchers: Brenda Clynch, Becky Cox
Illustrators: Ian Lycett, Catherine Ward
Indexer: Kay Ollerenshaw

Marshall Cavendish Corporation
Editorial director: Paul Bernabeo

Authors and Consultants

Dr. Roger Avery, BSc, PhD (University of Bristol)

Rob Cave, BA (University of Plymouth)

Fergus Collins, BA (University of Liverpool)

Dr. Julia J. Day, BSc (University of Bristol), PhD (University of London)

Tom Day, BA, MA (University of Cambridge), MSc (University of Southampton)

Bridget Giles, BA (University of London)

Leon Gray, BSc (University of London)

Tim Harris, BSc (University of Reading)

Richard Hoey, BSc, MPhil (University of Manchester), MSc (University of London)

Dr. Terry J. Holt, BSc, PhD (University of Liverpool)

Dr. Robert D. Houston, BA, MA (University of Oxford), PhD (University of Bristol)

Steve Hurley, BSc (University of London), MRes (University of York)

Tom Jackson, BSc (University of Bristol)

E. Vicky Jenkins, BSc (University of Edinburgh), MSc (University of Aberdeen)

Dr. Jamie McDonald, BSc (University of York), PhD (University of Birmingham)

Dr. Robbie A. McDonald, BSc (University of St. Andrews), PhD (University of Bristol)

Dr. James W. R. Martin, BSc (University of Leeds), PhD (University of Bristol)

Dr. Tabetha Newman, BSc, PhD (University of Bristol)

Dr. J. Pimenta, BSc (University of London), PhD (University of Bristol)

Dr. Kieren Pitts, BSc, MSc (University of Exeter), PhD (University of Bristol)

Dr. Stephen J. Rossiter, BSc (University of Sussex), PhD (University of Bristol)

Dr. Sugoto Roy, PhD (University of Bristol)

Dr. Adrian Seymour, BSc, PhD (University of Bristol)

Dr. Salma H. A. Shalla, BSc, MSc, PhD (Suez Canal University, Egypt)

Dr. S. Stefanni, PhD (University of Bristol)

Steve Swaby, BA (University of Exeter)

Matthew D. S. Turner, BA (University of Loughborough), FZSL (Fellow of the Zoological Society of London)

Alastair Ward, BSc (University of Glasgow), MRes (University of York)

Dr. Michael J. Weedon, BSc, MSc, PhD (University of Bristol)

Alwyne Wheeler, former Head of the Fish Section, Natural History Museum, London

Picture Credits

Heather Angel: 2922, 2933, 2990, 2998, 3014; **Ardea London Ltd:** Dennis Avon 2938, Jean-Paul Ferrero 3012, Kenneth W. Frink 2952, 2953, 3016, Ferrero Labat 2936, P. Morris 2902, 2934; **Neil Bowman:** 2919, 2927, 2975, 2979, 2980, 2981, 3003, 3010; **Bruce Coleman:** Franco Banfi 2915, 2917, Erwin & Peggy Bauer 2971, Erik Bjurstrom 2991, Mark N. Boulton 2951, Bob & Clara Calhoun 2985, John Cancalosi 2942, 2995, Bruce Coleman Ltd 2888, Alain Compost 2930, Christer Fredriksson 3000, Sir Jeremy Grayson 2929, Peter A. Hinchliffe 2973, Charles & Sandra Hood 2992, HPH Photography 2889, 2896, 2925, Johnny Johnson 2900, Gunter Kohler 2949, 2950, Wayne Lankinen 2891, Werner Layer 2987, Roine Magnusson 2965, Robert Maier 2946, 2947, Joe McDonald 2972, Dr. Scott Nielsen 2970, Tero Niemi 2892, Williams S. Paton 2926, 2943, Mary Plage 2986, Allan G. Potts 2967, 2978, 2982, Marie Read 2968, 2994, Hans Reinhard 2997, Kim Taylor 2894, 2928, Paul Van Gaalen 2977, Rinie Van Meurs 3004, Colin Varndell 2895, 2976, 2993, Jim Watt 2913, Gunter Ziesler 2937, J.P. Zwaenpoel 2931; **Corbis:** Patricia & Michael Fogden 3006, Stephen Frink 3007, 3008, Kevin Schafer 3018; **Chris Gomersall:** 2884, 2885, 2890; **NHPA:** A.N.T. 2916, 2961, 2962, 2984, A.N.T./Otto Rogge 2983, A.N.T./ Donald & Molly Trounson 3013, Daryl Balfour 2945, Anthony Bannister 3020, 3021, Mark Bowler 2969, N.A. Callow 2932, 2964, 2974, G.J. Cambridge 2956, Vicente Garcia Canseco 2944, Stephen Dalton 2903, Manfred Danegger 2893, Melvn Grey 2996, Martin Harvey 2906, 2907, 2948, Brian Hawkes 2898, Daniel Heuclin 2904, 2920, 2940, 2957, Stephen Krasemann 2914, Gerard Lacz 2958, 3015, Yves Lanceau 2935, Michael Leach 2960, Trevor McDonald 2918, Haroldo Palo Jr. 2908, 3019, Dr. Ivan Polunin 2921, Steve Robinson 2897, Jany Sauvanet 2988, Morten Strange 2923, 2924, Roger Tidman 2901, 2939, Dave Watts 2912, Bill Wood 2955, David Woodfall 3001, Norbert Wu 2954; **Oxford Scientific Films:** Jack Dermid 2966, Max Gibbs 2999, Tim Jackson 3017, Mark Jones 2887; **Still Pictures:** M.Boulton 3002, Mark Carwardine 2899, M. & C. Denis-Huot 3009, Michel Gunther 2959, David Hoffman 2905, Galen Rowell 2941, Roland Seitre 2963, Donald Tipton 2910, G. Verhaegen 3011, Norbert Wu 2911. **Artwork:** Ian Lycett 2886, Catherine Ward 2909, 2989, 3005.

Contents

WATER VOLE

The water vole's numbers are in decline in parts of Europe because of predation by mink and loss of suitable habitat. In Russia the species is also hunted by humans for its pelt.

The water vole is sometimes called the water rat. It is about the size of the common rat and can be easily mistaken for it in appearance. The head and body total 4¾–8½ inches (12–22 cm), and the tapering, hairy and ringed tail is 2¼–5 inches (6–13 cm) long. Its weight is in the range of 2½–4½ ounces (70–125 g) in winter to about 9 ounces (250 g) in summer. The female is slightly smaller than the male. The head is short and thick with a broad, rounded muzzle. The eyes are small and extremely shortsighted, and the small, round ears scarcely project beyond the surrounding fur. The limbs are relatively short. The nonwebbed feet have naked, pink soles with five rounded pads, but are covered with stiff hairs on the upper surface. All the toes bear claws. The thick, glossy fur varies from a blackish gray to a warm reddish brown above, sprinkled with gray, and the underparts are yellowish gray. A few melanistic (black) forms are found, as are albino strains.

The water vole, *Arvicola terrestris*, is found over most of Europe, parts of Russia and Siberia, Asia Minor, northern Syria, Israel and Iran. It is generally distributed throughout Britain, but its numbers are declining and it does not occur in Ireland or on the Scottish islands. The southwestern water vole, *A. sapidus*, lives entirely on land, burrowing rather like a mole. It is found only in France, Spain and Portugal.

Diving to safety

The water vole is found on the banks of streams, rivers and canals. It is thought to have a 4-hourly rhythm of activity throughout the day and night, with feeding periods of about half an hour alternating with periods of rest or random movement.

When a water vole suddenly plops into a stream or canal, its course can sometimes be tracked underwater. However, it often disappears immediately, to surface some distance away or retreat into a burrow in the bank, sometimes by an underwater entrance. It may regain the bank by an upper exit. It is a steady swimmer, its rate of progress being an even 2½–3 miles per hour (4–5 km/h), but it is less skillful in swimming than in diving.

Bedrooms and larders

The burrows made by water voles have long, winding passages with chambers for sleeping, lined with grass and hay, and chambers for storing food. The water vole digs them at great speed with the forefeet, throwing out the earth with the hind feet. It removes stones with the teeth and eats any roots that get in the way. The burrows sometimes cause damage to the banks of dikes and canals. However, the water vole also helps to keep waterways clear of weeds and rotting vegetation.

The water vole does not hibernate, but it has been reported to lay up considerable stores for when food is scarce. Although it is aquatic, steady rain will keep it in its burrow or cause it to gather food from near the mouth of the burrow and eat it inside. Like the water shrew (discussed elsewhere), the water vole is sometimes found in fields far from water. It marks its home range with a scent from glands on the flanks conveyed to the ground by the hind feet.

Vegetable eater

The water vole's diet features succulent grasses, flags, loosestrife, sedges and other plants that grow along river edges. The animal enjoys grains such as wheat, oats and millet, and apples are a special favorite. It is thought that it sometimes eats freshwater snails and mussels, as well as caddisworms and other insects, but this is not certain. Food stored in the burrow usually consists of different types of nuts, beech mast, acorns and the underground stems of horsetails.

WATER VOLES

CLASS **Mammalia**

ORDER **Rodentia**

FAMILY **Muridae**

GENUS AND SPECIES **Water vole, *Arvicola terrestris*; southwestern water vole, *A. sapidus***

ALTERNATIVE NAME
Water rat

WEIGHT
2½–9 oz. (70–250 g)

LENGTH
Head and body: 4¾–8½ in. (12–22 cm); tail: 2¼–5 in. (6–13 cm)

DISTINCTIVE FEATURES
Rounded body, no noticeable neck; pale brown to black body; long tail; small eyes

DIET
Aquatic vegetation, grass, roots, buds, twigs and fruit; perhaps some insects

BREEDING
Age at first breeding: about 2 months; breeding season: usually spring to early fall; gestation period: 20–22 days; number of young: 2 to 8; breeding interval: 1 to 4 litters per year

LIFE SPAN
About 6 months in the wild

HABITAT
***A. terrestris*: aquatic, burrowing in river, canal, pond and stream banks. *A. sapidus*: grassy cultivated habitats.**

DISTRIBUTION
***A. terrestris*: Europe, Russia, Siberia, Asia Minor, northern Syria, Israel and Iran. *A. sapidus*: France, Spain and Portugal.**

STATUS
***A. terrestris*: declining numbers. *A. sapidus*: near-threatened.**

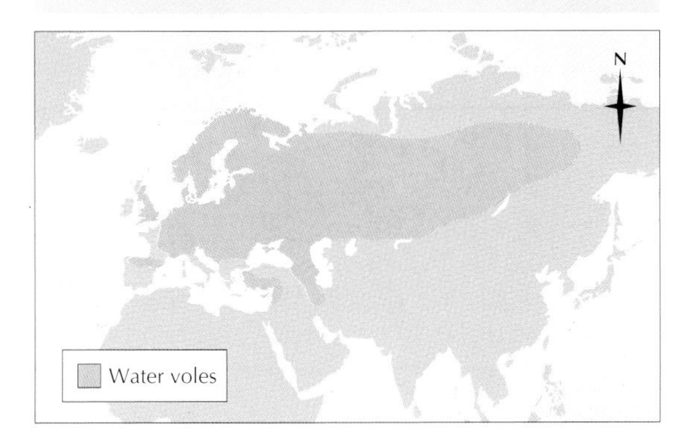

While the male water voles fight off potential predators, the females and babies usually dive into the water, returning to the bank after a short while to watch the proceedings.

Numerous litters

The breeding season is usually from early April to October. There may be fierce fighting among the males, but whether for possession of a female or for a territory is not clear. After mating, the female makes a thick-walled, globular nest of reeds and grasses in a chamber under the bank, in a hollow willow or in a disused bird's nest. Sometimes the male helps her make the nest. After a gestation period of 20–22 days, two to eight, usually five, naked and blind young are born. At 10 days old their eyes are fully open and they are covered with thick, smooth, golden-red hair. At 15 days they come out into the open but do not take easily to water. By the third week the young can feed themselves, swim fearlessly and are independent of the parents. The females can have up to four litters a year because the young of early litters mature so quickly, even breeding themselves before the winter. There are fewer litters if the summer is cold and wet.

The average life span is only about 6 months in the wild because older individuals are driven out of their territories by the younger voles and more readily fall prey to their numerous predators, such as herons, owls, otters, stoats, weasels, rats, pike, eels and large trout.

United in battle

Sometimes a whole family of water voles unites to fight off a predator. When a male member of a family is aware of an intruder approaching, it points with its snout upward and stands motionless, followed by all the other males. The females in the family and the babies retreat to the water's edge. One male then engages the predator, but if the predator is too strong for it, the water vole will run and join the females, and another male takes over the fight. The father is usually the last to enter the battle, but quite often by this time the predator is exhausted and escapes.

WATTLEBIRD

The huia, an extinct wattlebird, was most unusual in that the male (right) and female had bills of different lengths and shapes.

Also applied to Australian honeyeaters of the genus *Anthochaera* (discussed elsewhere), the name wattlebird, for the purposes of this account, refers to a small family of New Zealand birds, the Callaeidae. Of the three species, one is extinct and two are very rare, but efforts are being made to preserve the latter. It is uncertain which are the closest relatives of the wattlebirds, although they have been linked with the Australian mud-nesters, family Corcoracidae, and the birds of paradise and bowerbirds, family Paradisaeidae. It is probable that an ancestor crossed to New Zealand, where the three wattlebirds evolved. Wattlebirds have rounded wings, long tails, strong legs, long hind claws and wattles (fleshy facial parts), usually orange, at the corners of the mouth.

The kokako, *Callaeas cinerea*, also known as the wattled crow, is 17–18 inches (43–46 cm) long and jaylike, with a stout, curved bill. Its plumage is blue gray with a velvet-black strip in front of the eyes and above the bill. Fewer than 400 pairs are in existence. The North Island form, *C. c. wilsoni*, has bright blue wattles, while the South Island form, *C. c. cinerea*, which lived on Stewart Island but may now be extinct, has orange wattles with blue at the root. The saddleback, or tieke, *Philesturnus carunculatus*, is 10 inches (25 cm) long, with glossy black plumage except for a chestnut saddle over the back. Like the kokako, the saddleback occurs in a South Island form, *P. c. carunculatus*, and in a North Island subspecies, *P. c. rufusater*.

Europeans encountered the now-extinct huia, *Heteralocha acutirostris*, only in the southern part of North Island. The remaining species were once widespread in both islands. The saddleback is now restricted to Hen Island in the north, the South Cape Islands in the south and a few more islands where it has been introduced in recent years in an attempt to preserve it. The kokako occurs in a number of areas on North Island but is rare and becoming rarer.

Threatened by rats

Wattlebirds are very active but rarely fly. They spend most of their time searching for food on the floor of New Zealand's primeval forests, and the felling of these forests is one of the main reasons for the wattlebirds' decline. Another important factor is the introduction of nonnative predators. The saddleback now survives only where rats and other mammalian predators have failed to spread. In the early 1960s, ship rats came ashore at the South Cape Islands, threatening the South Island subspecies. A rescue operation was mounted to save the threatened saddlebacks, and in 1964, 36 birds were captured and flown to other islands. The current population of the South Island subspecies is about 650 birds on 11 small, predator-free islands. In the north, some North Island saddlebacks have been moved from Hen Island, and this form is now found on nine large islands.

Fruit, leaves and insect food

The kokako feeds mainly on fruit and leaves but also eats small ground-living invertebrates. The latter form the main part of the saddlback's diet, and the saddleback follows bands of other small birds, apparently to catch the insects they disturb. Wattlebirds sometimes hold their food in one foot while tearing it with the bill.

The saddleback probably mates for life. Outside the breeding season it lives in flocks, but each pair takes up a territory in spring. The male

KOKAKO

CLASS **Aves**

ORDER **Passeriformes**

FAMILY **Callaeidae**

GENUS AND SPECIES ***Callaeas cinerea***

ALTERNATIVE NAMES
Bellbird; organbird

WEIGHT
8–8½ oz. (230–240 g)

LENGTH
Head to tail: 20 in. (50 cm)

DISTINCTIVE FEATURES
Blue gray with velvet-black face mask; longish tail. North Island form: ultramarine wattles. South Island form: orange wattles.

DIET
Fruits; insects, nectar, leaves; fern fronds

BREEDING
Age at first breeding: probably 2 years; breeding season: November–December; number of eggs: 2 or 3 (usually); incubation period: about 25 days; fledging period: 27–28 days; breeding interval: 1 year

LIFE SPAN
Not known

HABITAT
Temperate forest

DISTRIBUTION
New Zealand

STATUS
North Island form: very rare. South Island form: probably extinct.

displays to the female frequently, bowing, singing and inflating his wattles. The wattles are larger in the male than in the female and reach their greatest size during the breeding season.

The saddleback's nest is a large, loosely woven cup of twigs, leaves and ferns, often placed in a hollow tree or rocky crevice or among dense foliage on the ground. The two or three eggs are incubated by the female for 3 weeks or longer. Both parents feed the young. Young North Island saddlebacks look like the adults, but South Island saddlebacks have a dark brown plumage for 1 year. At one time these young birds were thought to be a separate species and were called jackbirds.

The kokako is limited to virgin or only barely altered forests in New Zealand. It is sometimes locally called the bellbird due to its far-carrying, bell-like song.

The loss of the huia

When Europeans arrived in New Zealand, the huia already had a restricted distribution in the mountainous regions at the southern end of North Island. The Maoris snared them, luring them by imitating their calls, to obtain the tail feathers, which were prized as ornaments to be worn in the hair. The Maoris had been hunting huias for many years when the Europeans began to collect the birds for museums. It is unlikely that either of these two kinds of collecting seriously affected huia numbers, and introduced predators and forest clearance are probably to blame for the species' disappearance. The huia was last recorded in 1907.

The loss of the huia is particularly regrettable because of a unique distinction that existed between the sexes in this species. There are numerous examples in this encyclopedia of the adaptation of bills to accommodate specific feeding habits, but there is no other example of two kinds of bill within one species. The male huia had a stout, relatively straight bill with which it chiseled insects and their grubs out of wood. The long, curved and rather pliable bill of the female was for probing holes and crevices in wood. In the past, common belief held that the two sexes worked together in feeding, but there is no evidence of this taking place.

WAXBILL

Waxbills make up for their plain singing by their bold coloration. The orange-cheeked waxbill,* E. melpoda*, has a characteristic bright red bill.

THE WAXBILLS ARE A GROUP of about 15 small, colorful seed-eating species that are popular cage birds. Waxbills are related to the sparrows and weavers, and the waxbill subfamily, Estrildinae, includes the mannikins, munias, cordon-bleus, silvereyes and many others familiar to bird fanciers. Unfortunately, several have different common names, which makes the term waxbill open to confusion. The cordon-bleus, for instance, are also called blue waxbills. The waxbills proper belong to the genus *Estrilda*, which also includes the striking red and green avadavats (described elsewhere).

Waxbills are small, around 4 inches (10 cm) long, and many have finely barred upperparts. The species *E. astrild*, known as the common waxbill, is often known to cage-bird enthusiasts as the St. Helena waxbill. It is brown with fine barring, and there is a scarlet patch around the eye. The cheeks and throat are white, and in the male there is a pink tinge to the underparts. This bird is found in many parts of Africa and has been introduced into St. Helena and Brazil.

Other waxbills have a similar confusion of names. The gray, red-eared or black-rumped waxbill, *E. troglodytes*, is also called the common waxbill. The upperparts are grayish brown with a pink tinge, and the underparts are pale gray with a pink tinge that turns to crimson on the belly. There is a crimson stripe through the eye and the rump is black. The gray waxbill became established in Portugal from aviary escapes during the mid-20th century.

One of the smallest waxbills is the 3½-inch (9-cm) locust finch, *E. locustella* (also known as *Ortygospiza locustella*), which flies in dense swarms. Its plumage is almost black, with red on the face and throat. The smallest waxbill of all is the zebra or orange-breasted waxbill, *Amandava subflava*, which has a crimson streak through the eye and a crimson rump. The throat is yellow, becoming scarlet underneath, and the sides are barred with yellow. With the exception of the red and green avadavats in Asia, waxbills live in sub-Saharan Africa.

Grain eaters

Outside the breeding season waxbills are gregarious, living in parties, some of only a few birds, but others, as in the locust bird, comprising large flocks. The members of a party continually call to each other with shrill or soft monosyllables intended to inform each waxbill of its fellow's position and to keep the party together. Waxbills are found mainly near rivers or in swampy country, where they feed on seeds, particularly those of grasses; they are particularly abundant in grassland and in crops of cereals, in association with other seed-eaters, such as mannikins and widowbirds. In Sierra Leone the flocks are followed by rats, which feed on the seeds they spill. In general, waxbills occur in too few numbers to be pests. They also eat some insects and catch flying termites.

Nests woven from grass

The typical waxbills differ from their near relatives by building nests with tubular entrances projecting from a ball of grass that are very much like the nests of sparrows and weavers. The nest is built of grass stems or flowering heads woven into an untidy mass and fastened to vertical stems or placed on the ground among grass or herbage. Some waxbills decorate the nest with paper, damp soil, feathers and other materials.

A peculiar feature of the nests of true waxbills is that there is a so-called cock nest incorporated into the top or side of the nest or built a short distance away. It has been said that the cock nest is used as a roost by the member of the pair that is not incubating the eggs. There is, however, no scientific proof of this, and some

COMMON WAXBILL

CLASS **Aves**

ORDER **Passeriformes**

FAMILY **Estrildidae**

SUBFAMILY **Estrildinae**

GENUS AND SPECIES ***Estrilda astrild***

ALTERNATIVE NAME
St. Helena waxbill

LENGTH
Head to tail: 4½ in. (11.5 cm); wingspan: 4¾–5½ in. (12–14 cm)

DISTINCTIVE FEATURES
Small, compact finch; strong, conical, red bill; grayish brown plumage with fine barring; scarlet patch around eye; fairly long tail; male has pinkish wash to underparts

DIET
Mainly seeds and grain; also termites and other flying insects

BREEDING
Age at first breeding: 6–12 months; breeding season: September–January (southern Africa); number of eggs: usually 4 to 6; incubation period: 11–12 days; fledging period: 17–21 days; breeding interval: several broods per year

LIFE SPAN
Up to 4 years

HABITAT
Grassland, often near rivers; also swamps, reed beds, gardens and outskirts of human habitations

DISTRIBUTION
Much of sub-Saharan Africa, south from Cameroon and southern Sudan; introduced to Brazil, St. Helena and Hawaii

STATUS
Common

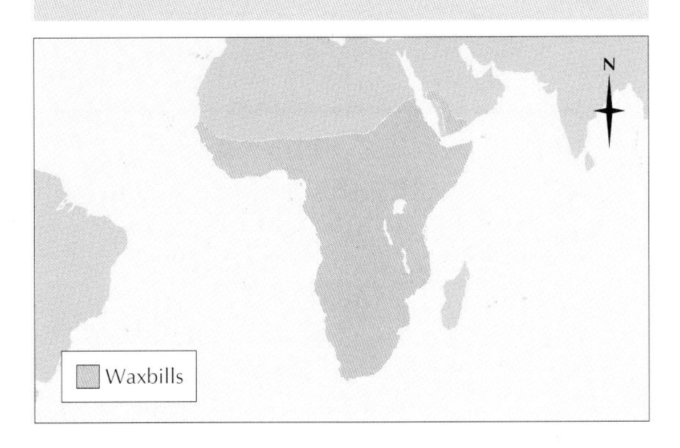

A male blue waxbill,* Uraeginthus angolensis, *in the Traansvaal, South Africa. This species is commonly seen among savanna thornbushes.

ornithologists have suggested the cock nest may mislead predatory birds into overlooking the real nest. The nest is built by the female waxbill, but the male helps with the decoration and with lining the nest with feathers. Both sexes incubate the four to six white eggs, which hatch in 11–12 days. One species feeds the chicks by regurgitating seeds. The chicks beg for food by gripping the parents' bills. The young can fly in 17–21 days.

Getting their own back?

Many waxbills are parasitized by some of the related whydahs, also known as widowbirds. One species that does this is the pin-tailed whydah, *Vidua macroura*. The whydahs lay their eggs in the waxbills' nests, and their young are brought up with the young waxbills. Not all the whydahs are, however, parasites, and one waxbill, the zebra waxbill, has to a certain extent reversed the situation: it lays its eggs in the nests of whydahs and bishop birds, but only when they have been abandoned. Whydahs and bishop birds finish nesting in March, and their vacant nests are adopted and relined by the waxbills, which then start their nesting season.

WAXWING

The three species of waxwings are named after the red tips of their secondary flight feathers, which resemble blobs of sealing wax. Similar but smaller blobs are also found on the tail feathers. The waxwings include the cedar waxwing, *Bombycilla cedrorum*, the Bohemian waxwing, *B. garrulus,* and the Japanese waxwing, *B. japonica.*

The waxwings are starling-sized birds, 7 inches (18 cm) long, with prominent pointed crests, fairly long wings and slightly rounded tails. The bill is short and slightly hooked. The nearest relatives of the waxwings are the silky flycatchers of the United States, such as the crested phainopepla, *Phainopepla nitens.*

The Bohemian waxwing breeds from northern Scandinavia to Kamchatka in Siberia. It also breeds in western Canada and Alaska. This bird is generally vinaceous- (wine-) brown in color with a bold crest on the head, and there is black around the eyes and chin. The secondary flight feathers have bright red waxy tips and the primary flight feathers are black with yellow and white tips. The black tail is tipped with yellow.

The cedar waxwing of southern Canada and the northern United States is very similar but the plumage is generally paler and it lacks the yellow and white on the wings. The Japanese waxwing, *B. japonica,* resembles the cedar waxwing but with a red tip to the tail, red bars on the wings and no waxy red droplets. It lives in eastern Siberia and migrates to China and Japan.

Irregular migrations

Apart from their appearance, waxwings are noted for their irregular migrations and wanderings. Small numbers of waxwings usually migrate south in the winter, but there are occasional mass movements of large numbers when waxwings can be seen in flocks. On rare occasions, Bohemian waxwings fly as far south as the Mediterranean. By comparison, cedar waxwings are rather more regular visitors to Central America.

Many observers have commented that the waxwings in these winter flocks are sluggish, perching motionless for long periods and allowing themselves to be approached quite closely.

Waxwings do not generally exhibit territorial behavior, although they do defend their nests. Pictured is a Bohemian waxwing.

BOHEMIAN WAXWING

CLASS **Aves**

ORDER **Passeriformes**

FAMILY **Bombycillidae**

GENUS AND SPECIES ***Bombycilla garrulus***

WEIGHT
1¾–2½ oz. (50–75 g)

LENGTH
Head to tail: 7 in. (18 cm)

DISTINCTIVE FEATURES
Starling-sized bird; generally vinaceous- (wine-) brown color; bold crest on head; bright red, waxy tips to secondary flight feathers; yellow and white tips to primary flight feathers; black bib; black tail with yellow tip; long wings; slightly rounded tail

DIET
Mostly insects in summer; mostly fruit and berries in winter

BREEDING
Age at first breeding: 1 year; breeding season: late May–June; number of eggs: 5 or 6; incubation period: 14–15 days; fledging period: 14–15 days; breeding interval: 1 year

LIFE SPAN
Not known

HABITAT
Breeds in taiga forest, especially spruce and pine; nests preferentially in stunted conifers

DISTRIBUTION
Taiga zone from Norway in west to Kamchatka, Siberia, in east; also in Alaska and western Canada

STATUS
Common

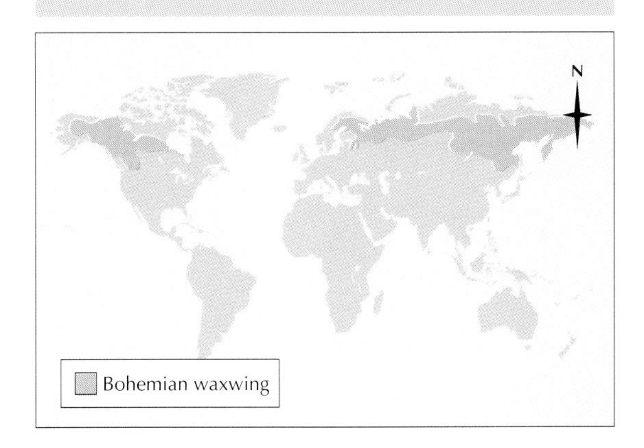

Cedar waxwings, such as the one above, sometimes pass food from one to another even when they are not actively involved in courtship.

When they are perching, the members of a flock often huddle closely together. They sometimes touch each other, although they usually keep an inch (2.5 cm) or so apart.

Even in winter, waxwings sometimes feed each other, as they do during courtship, when food, or occasionally an inedible object, is passed to and fro between two birds. A line of cedar waxwings have been observed to pass food from one to another.

The nomadic habits of the waxwings extend even to their breeding. Waxwings nest in the coniferous and birch forests of northern Europe, Asia and America. However, they shift their breeding grounds from year to year, depending partly on the local abundance of food. The birds' general lack of territorial behavior is perhaps related to this. Nesting waxwings defend no territory other than the nest itself, and the song is very poor, being no more than a thin trill.

Berry eaters

In the summer, waxwings eat mainly insects, catching flies in the air and foraging on the ground. They also eat flower petals and oozing sap but the main food throughout the year is berries such as those of juniper, yew, rowan and elder. Waxwings also eat blackberries, hawthorn, holly, cherries and many other berries.

When the food supply is plentiful, waxwings, such as this Bohemian waxwing, overwinter in the north of their range. When there is little food they must fly south in large numbers.

Limited rivalry

Waxwings' nests are usually solitary but, as these birds do not defend their territory, the nests are sometimes placed close to each other. The nest is a cup of twigs lined with moss and grass and built in a pine or birch tree. The Bohemian waxwing prefers to nest in stunted conifers.

Even at the start of the nesting season waxwings are fairly sociable, and there is only a restricted amount of rivalry between the males. During courtship, the feeding of the female by the male is accompanied by a display in which both birds puff out their feathers, particularly those on the rump.

The male waxwing also feeds the female while she incubates the five or six eggs for 14–15 days. The male plays a small part in incubation and both parents feed the chicks.

Starvation exodus

Provided there is plenty of food, many waxwings spend the winter in the northern parts of their range, even north of the Arctic Circle. Others migrate southward, and every few years there is a mass exodus, called an irruption. In winters when an irruption has taken place, waxwings are seen in large flocks, and stragglers appear far to the south of their usual limit of winter migration. In the winter of 1965–1966, there was an irruption of waxwings from northern Scandinavia, Finland and northwestern Russia, and the birds appeared as far south as Portugal, Sicily and Greece.

Irruptions of crossbills, lemmings (both discussed elsewhere in this encyclopedia), waxwings and other animals are usually caused by a season of plentiful food, when the bird population expands, followed by a failure of the food supply, which then forces the large population to travel in search of food. The waxwing irruption of 1965–1966 was predicted in advance because these conditions had been fulfilled. The winter of 1964–1965 had been unusually mild and there had been a good crop of rowan berries in northeastern Europe. As a result, the waxwings survived the winter well and large numbers were able to nest in 1965.

Rowans, like other plants, cannot fruit well for two years running. In the fall of 1965, the large population of waxwings was confronted with a food shortage and they were forced to move on. They left the White Sea region in Russia in late September and arrived in southern England in mid-November, with stragglers reaching Portugal and Sicily in January.

WEASEL

Although similar in form to its near relative the stoat, *Mustela erminea*, the weasel is smaller and lacks the black tip to the tail. The most common species is the least weasel, *M. nivalis*, in which the average head-and-body length of an adult male is 8½ inches (22 cm), plus a tail of 2¾ inches (7 cm). The female is an inch (2.5 cm) or so shorter and usually weighs 2 ounces (57 g) against the male's 4 ounces (113 g). Because of their smaller size, the females were formerly believed to represent a distinct species. The long, slender body, short limbs, long neck and small head give the weasel a snakelike profile, which is heightened by its gliding movements. The fur is reddish brown with white on the throat and underparts, but the line of demarcation between the colors is less pronounced than in the stoat. It has been said that a weasel is small enough to pass through a wedding ring.

The least weasel is found across northern North America. It occurs across Europe and Asia to Japan, and from Siberia south into China and Afghanistan. Its range extends into North Africa, and it has been introduced into New Zealand, the Azores and a number of other small islands. There are several other species of weasels, found from Canada to South America and from Siberia to the Himalayas, southern China and Southeast Asia, but this article focuses on *M. nivalis*.

Like stoats, weasels undergo a change to a white coat in the fall in the more northerly parts of their range, although as a rule there is no seasonal change in their fur in more southerly parts; an occasional individual may, however, be white or partially white in winter.

Ferocious killer

The weasel is found in almost every type of habitat including woods, scrubland, hedgerows, rocky country, barns and even in large towns. In captivity the weasel is mainly nocturnal. In the wild, however, it hunts by day and night, alternating active bouts of 10–45 minutes with rest periods, some of which may last 3 or 4 hours. It is swift and agile in movement, a good climber and swimmer and a relentless killer.

Like other members of its family, the weasel is bold and ferocious out of all proportion to its size and readily attacks animals larger than itself. It has been seen struggling to haul along a nearly

Normally carnivorous, the highly active weasel cannot survive without food for more than 24 hours. Any energy source is considered at times of hunger, including red currants.

A weasel carries off a yellow-necked mouse, Apodemus flavicollis, *to a cache to eat later. A weasel usually eats 5 to 10 meals, each of 1/18–1/9 oz. (2–4 g), in a 24-hour period.*

full-grown rat, two or three times its own weight, killed by a trademark bite through the base of the skull. Sometimes weasels hunt in pairs or in family parties. The normal method of hunting is to stalk or trail the prey mainly by scent, but also by sound and sight, and then to pounce swiftly and kill with a bite on the back of the victim's head, puncturing the braincase.

Staple diet of voles and mice

The weasel's food includes rats, mice, voles, moles, frogs, small birds and their eggs and an occasional fish. It swims after water voles and climbs trees and bushes to rob birds' nests and bird boxes of eggs or young. When it is very hungry, it eats shrews or carrion, and does some damage in poultry runs. Voles and mice are, however, the main prey, a weasel's small size enabling it to follow them in their underground runs, although males find these rather a squeeze and rely also on mole runs. So dependent is the weasel on these rodents that their population fluctuations are mirrored in its own fortunes. For example, when the viral disease myxomatosis killed many rabbits in Britain in the 1950s, the ungrazed hedgerows became overgrown and provided new homes for small rodents. These flourished, and the weasel flourished with them. By contrast, the stoat, which had depended largely on the rabbit for prey, declined for some years.

Weasels occupy territories that they mark with musk from the glands under the tail. This musk is also released when an animal is severely disturbed, and it may also be used to bring the

LEAST WEASEL

CLASS **Mammalia**

ORDER **Carnivora**

FAMILY **Mustelidae**

GENUS AND SPECIES ***Mustela nivalis***

ALTERNATIVE NAME
Cane weasel

WEIGHT
5/7–7 oz. (25–250 g)

LENGTH
Head and body: 4½–10½ in. (11–26 cm); tail: ¾–3⅕ in. (2–8 cm)

DISTINCTIVE FEATURES
World's smallest carnivore, with slender body, short legs and long neck; northern populations molt from summer coat of chestnut-brown back and white belly to winter coat of white

DIET
Small rodents, including voles, mice and rats; shrews, rabbits; birds and eggs; fish, amphibians; carrion

BREEDING
Age at first breeding: 3–4 months; breeding season: may be year-round, usually spring to late summer; gestation period: 34–47 days; number of young: 3 to 10; breeding interval: female may have 2 litters each year

LIFE SPAN
10 years in captivity; much less in wild

HABITAT
Highly variable, including agricultural land, forest, tundra and large towns

DISTRIBUTION
Native to Northern Hemisphere; introduced into New Zealand, Azores and elsewhere

STATUS
Globally widespread but generally rare within range

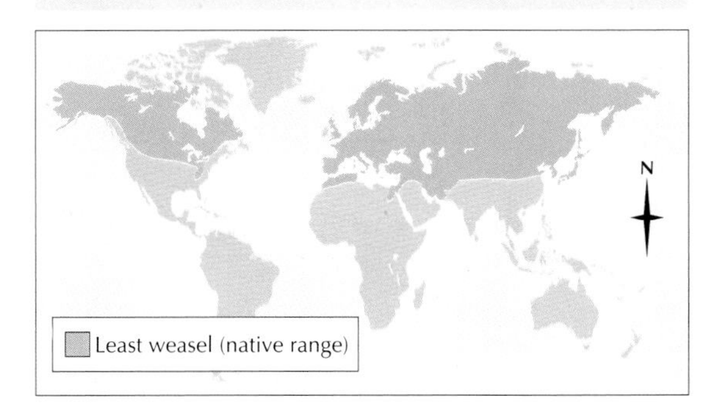

males and females together in the breeding season. The voice is a guttural hiss when alarmed or a short screaming bark when disturbed.

Two litters in a season

The female weasel builds her nest of dry leaves, grass or moss in a hole in a bank or low down in a hollow tree. Pregnancy usually occurs in any month from March to August but is most frequent in April and May. There is no delayed implantation, and the gestation period is about 6 weeks. There may be two litters in a season, especially when the vole population is healthy. Each litter consists of three to eight, but usually five, kittens. They are weaned at 4–5 weeks and are taught by their mother to hunt and kill. Young males of the first litter grow rapidly and are sexually mature by August, as are some of the females. Second litters grow more slowly and do not mature until their second year.

In years when prey numbers and spring temperatures are both low, a female weasel may rear no young at all.

Natural mortality

The weasel's natural predators are the larger hawks, owls, foxes, wild and domestic cats and sometimes stoats. The numbers taken, however, are not large and the effect on the weasel population is negligible. A much more important cause of mortality is the climate. Its long, slender shape and short fur render the weasel vulnerable to heat loss, and it can survive the highly stressful conditions of winter only when it finds shelter in a fur-lined nest, often under an insulating blanket of snow.

Hunted as a pest

In former times, every gamekeeper and farmer considered the weasel a dangerous pest for its alleged raids on game birds and poultry. The weasel was shot and trapped extensively. The accusations of poultry and game killing were undoubtedly justified to some extent, but today many people have realized that the weasel does more good than harm by keeping down the numbers of small rodents in the countryside. It has been estimated that a male weasel kills at least 500 small rodents a year, while a female accounts for about 300. This, coupled with the fact that there has been a steady decline in widespread game preserves in recent years, is reducing the number of weasels killed each year. Despite this, many weasels are still taken by gamekeepers each year, especially in Britain.

Male weasels spend more time above ground than females. A study in Scotland found males spent 50 percent of active time below matted grass, compared to 90 percent in females.

WEAVER

A cape weaver, Ploceus capensis, *lays the foundations of its nest. About 6¾ inches (17 cm) in length, this species occurs in moist areas of South Africa.*

THE WEAVERS AND THEIR allies make up the family Ploceidae. There are 114 species in this family, including the buffalo-weavers, sparrow-weavers, queleas, bishops and widowbirds. These birds all have important features in common, but this account will concentrate on birds in five genera. Most true weavers are in the genus *Ploceus*, but related closely to them are the malimbes in the genus *Malimbus* and three single-species genera, *Pachyphantes, Anaplectes* and *Brachycope*. Most of the other main groups in the family Ploceidae have been dealt with separately elsewhere.

These five genera of weavers contain about 73 species. Most live in Africa south of the Sahara, but a few live in southern Asia, and the village weaver, *Ploceus cucullatus*, also known as the black-headed weaver, has been introduced into Haiti. Weavers are sparrow-sized birds, about 6 inches (15 cm) in length, and have the conical, seed-eating bill of the house sparrow, *Passer domesticus*. The males of many weavers have bright plumage during the breeding season but revert to the same drab, streaky plumage as the females outside the breeding season. The males of the 11 malimbe species have red in their plumage. The red-headed malimbe, *Malimbus rubricollis*, is black all over with a red cap on the crown and nape. The males of the *Ploceus* species have yellow plumage. The village weaver is one of the best known of these. It is golden yellow with a black head and black streaks on the back and wings. The five Asian *Ploceus* species are much alike. The baya weaver, *P. philippinus*, which ranges from Pakistan and Sri Lanka to Sumatra, is mainly black above and pale brown underneath, with yellow on the head and neck above the eye and black beneath it.

Varied habitat

Weavers are found, often in large flocks, in a variety of habitats but always where there are trees in which they can roost and nest. The malimbes live in evergreen forest, whereas the *Ploceus* weavers occur in wooded country of various types. For instance, the masked weaver, *P. intermedius*, prefers watercourses, and often nests on branches that overhang water. Other *Ploceus* species are found in dry savannas or in marshes. Weaver species in which the males are brightly colored in the breeding season generally live in drier areas. The dull eclipse plumage is

SPECTACLED WEAVER

CLASS **Aves**

ORDER **Passeriformes**

FAMILY **Ploceidae**

GENUS AND SPECIES ***Ploceus ocularis***

WEIGHT
Nearly 1 oz. (25–26 g)

LENGTH
Head to tail: 6–6¼ in. (15–16 cm)

DISTINCTIVE FEATURES
Adult male: sharp, pointed black bill; yellow eye; golden-yellow crown and underparts, shading to orange around black spectaclelike eye patch and black throat; remainder of upperparts yellow green; remainder of underparts bright yellow. Adult female: similar to male but throat golden yellow. Juvenile: similar to female but bill brown.

DIET
Insects, spiders, millipedes; nectar, fruit and seeds

BREEDING
Age at first breeding: 1 year; breeding season: September–February; number of eggs: 2 to 4; incubation period: 13–14 days; fledging period: 18–19 days; breeding interval: 1 year

LIFE SPAN
Not known

HABITAT
Riverine forest, forest edge, dense scrub; gardens and parks

DISTRIBUTION
Much of sub-Saharan Africa, from Guinea in west to Ethiopia in east and South Africa in south; avoids very arid regions

STATUS
Fairly common

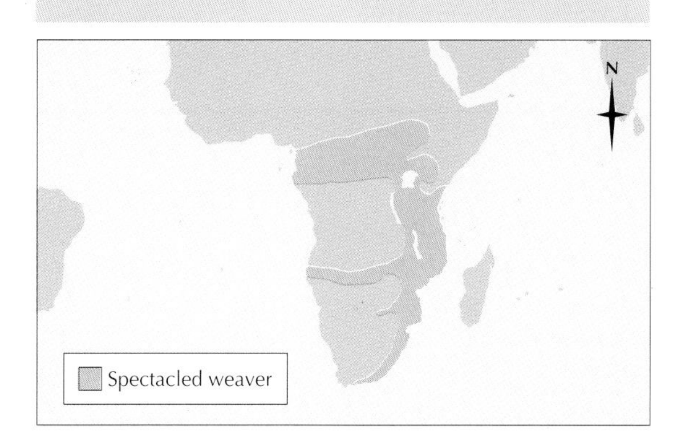

This southern masked weaver,* Ploceus velatus, *has nearly completed its nest. This weaver is a group-nesting species and, like the masked weaver, often builds in trees over water.

associated with the formation of large flocks that wander about the country in the dry season to search for food. Whereas these weavers are gregarious, some species, such as the spectacled weaver, *P. ocularis,* occur singly or in pairs.

Grain, seeds and insect food

The majority of weavers feed on seeds, particularly grass seeds, and several species have become pests of grain crops, although not to the same extent as the red-billed quelea, *Quelea quelea,* which is also known as the Sudan dioch. Insects are also eaten, particularly by a few *Ploceus* species that have slender bills and by the malimbes. These weavers hunt among the foliage or along branches and on tree trunks, agilely hopping up or down.

Numerous species of weavers are gregarious and take over trees with communities of nests.

Well-protected nests

Weavers are named after the elaborate flask-shaped nests that they make from strands of grass or palm fronds. Each nest is separate, but a large tree may contain many nests. In the case of the village weaver, it may hold hundreds of them. The nest is built by the male, which cuts a notch in a palm frond or grass stem then strips a 2-foot (60-cm) thread from it, behavior that can make it a pest in palm plantations. The strips are first woven into a loop, which acts as a foundation for a hollow, ball-shaped nest. An entrance tunnel up to 2 feet long is sometimes attached.

When the main structure is completed, the weaver displays at the entrance to attract a female. If she accepts both male and nest, mating takes place. The female lines the nest and lays two to four eggs, which she alone incubates. In many species, particularly those in which the male has bright breeding plumage, the male builds several nests, courting and installing a female in each. When they run out of partners, the males continue to build nests, which remain unused and usually half-finished. Some species of weavers are monogamous, and the insect-eaters are territorial.

It must be extremely difficult for any predator to invade a weaver's nest. The structure is not only suspended from a twig but also may be protected by a vertical tunnel. It is hard to imagine a snake or a mongoose, for example, managing to climb down the outside of the nest and then turn up through the entrance. There have been cases of parasitic birds that lay their eggs in weaver nests becoming stuck in these entrance tunnels. Even so, weavers very often employ a second line of defense by building their nests near aggressive animals that may dissuade other predators. The village weaver, for instance, often nests near human habitations or near the nests of large birds of prey, and in Peninsular Malaysia the baya weaver builds its nests in trees swarming with ants.

Variable-pattern eggs

Although they are well protected against many intruders, weaver nests are not invulnerable to cuckoos. The females of some cuckoos, including the European cuckoo, *Cuculus canorus*, can mimic the eggs of their hosts, even though the individual cuckoos do not all parasitize the same species. Parasites of weavers, however, have an additional problem: weaver eggs are very variable, those of the masked weaver having different patterns in different localities. Remarkably, the Didric cuckoo, *Chrysococcyx caprius*, follows suit and has the ability to mimic local patterns, laying eggs that match the different weaver eggs.

WEDDELL SEAL

Of the four Antarctic seals, the Ross, crabeater, leopard and Weddell, most is known about the Weddell seal, *Leptonychotes weddelli*. Unlike the others, this species breeds around the coastline, and because it lives under the sea ice near land it is possible to walk out over its home and to study it with comparative ease. When first discovered, this seal was called Weddell's sea-leopard, after its 19th-century discoverer Captain James Weddell and its spotted coat. The Weddell seal could be mistaken for a leopard seal but for its small head and distinctly tubby body. Adult males grow to 8–9 feet (2.4–2.7 m) and females grow a little longer, up to 11 feet (3.3 m). The color of the coat varies, being usually dark or light gray with white streaks and spots. In summer the fur fades to a dirty brown. The seal has a doglike face with rolls of fat around the neck and long, curling whiskers.

Weddell seals are found all around the coasts of Antarctica and its neighboring islands. The most northerly breeding colony is on South Georgia, but sometimes Weddell seals are seen around the Falkland Islands, New Zealand and southern Australia.

The Weddell seal is one of the largest seal species. It lives farther south than any other seal species.

Living icebreaker

The Weddell seal is the most southerly living mammal, being found in considerable numbers along the coasts of the Weddell and Ross Seas. Unlike many other seals, the Weddell seal prefers the fast ice of the Antarctic ice shelf instead of drifting pack ice. It spends most of its time in the water, where the temperature is usually higher than the air temperature, but on fine days in particular the seal hauls itself out of the water to bask. It prefers to lie on the ice, but if this is not available it chooses the smoothest rocks it can find. It probably hauls out to digest after feeding because while it is diving its blood is diverted from the intestine to the essential organs, such as the brain.

Throughout the winter in most parts of the Weddell seal's range, and over the whole year in some parts, the sea is frozen to a depth of 4 feet (1.2 m) or more. To breathe the seals either have to find a crack or lead in the ice or they must carve a breathing hole. These breathing holes are opened, and kept open, by vigorous sawing actions of the mouth. As a result, the teeth of Weddell seals are blunted, and the loss of the teeth may be a prime cause of death in old seals. It is known that Weddell seals travel up to 1⅕ miles (3 km) from one breathing hole to another, but these animals are not migratory. How the seals locate their breathing holes remains a mystery.

Diving and fishing

Weddell seals eat mainly fish, such as Antarctic cod and icefish, which they find on the seabed and in midwater. Fish of up to 45 pounds (20 kg) have been found in their stomachs. Weddell seals also eat squid and octopus and many kinds of planktonic crustaceans. Krill are usually eaten only when the seals are hunting in the pack ice. Like other seals, young Weddell seals start by eating only open-water prey, such as crustaceans, presumably because their ability to dive has not yet developed sufficiently. Only the adults can reach deepwater prey, such as cod.

Chilly reception

For most of the year Weddell seals are solitary, but from August or September (the southern early spring) females start to haul out through cracks and holes in the ice to give birth to their pups. They may gather in groups of 20 or more at this time, but these concentrations are mainly the result of the seals taking advantage of the available openings in the ice. The pregnant seals do not emerge more than about 300 feet (90 m) from the shore unless there is a suitable rock or islet offshore. In the northern parts of their range Weddell seals sometimes give birth on land.

By the time a female has finished nursing her pup it weighs around 250 lb. (113 kg), and its sleek, blubbery body is sufficiently developed to permit dives of up to 300 ft. (90 m). At this point the mother is ready to leave her pup to fend for itself.

The pups, usually two in number, are just under 5 feet (1.5 m) long, and are born a few days after the mother has hauled out on the ice. Born into an intensely cold world, they have no protective layer of blubber and may suffer a drop in temperature of over 100° F (38° C) as they slip from the mother's body and hit the ice.

The cow (female) stays with her pups until they are weaned 6–7 weeks later. During this time she does not feed and changes from a plump animal hardly able to heave herself over the ice to a skinny creature whose bones are visible beneath the skin. She loses about 300 pounds (136 kg), much of which is passed to the pups as milk and converted into fat, so that by the time they are weaned the pups can hardly move. The pups may first enter the water when only a few days old. The mother is very solicitous and even helps them out of the water.

The males do not help rear the pups. During the pupping period they establish territories and occasionally they fight. These fights have been witnessed on the ice. The seals appear to be ferocious, but the tough hide, thick blubber and blunt teeth prevent much serious damage from being done, although male Weddell seals are often found with one eye blinded or their flippers mutilated. Mating takes place after the pups leave the mothers. It has never been seen, so presumably mating takes place underwater.

Studying the seals

Although Weddell seals live in an inhospitable part of the world, they are quite easy to study. They do not fear humans, and sometimes it is possible to tag a hind flipper while the seal just looks on placidly. Their frozen habitat means scientists can walk out to them, and scuba-diving scientists can examine their aquatic habitat, recording the territorial, birdlike trills and whistles that can be heard from above the ice.

WEDDELL SEAL

CLASS **Mammalia**

ORDER **Pinnipedia**

FAMILY **Phocidae**

GENUS AND SPECIES ***Leptonychotes weddelli***

WEIGHT
880–1,320 lb. (400–600 kg)

LENGTH
Male: 8–9 ft. (2.4–2.7 m); female: up to 11 ft. (3.3 m)

DISTINCTIVE FEATURES
Plump body; large, strong, forward-pointing canine teeth; coloration variable, generally dark or pale gray coat spotted with white streaks and spots; dirty brown fur in summer; short neck; small head; fat rolls around neck; long, curling whiskers

DIET
Cod, silverfish, squid, octopus, krill

BREEDING
Age at first breeding: 3–6 years (female), 7–8 years (male); breeding season: birthing August–October, mating again soon afterward, sometimes as late as December; gestation period: 21–22 weeks; number of young: usually 2 (1 to 4); breeding interval: 1 year

LIFE SPAN
Up to 22 years

HABITAT
Fast ice, not drifting pack ice, and open sea down to 2,000 ft. (600 m)

DISTRIBUTION
Antarctic ice shelf and Antarctic islands, including South Sandwich, South Shetland, South Georgia and South Orkneys

STATUS
Not threatened

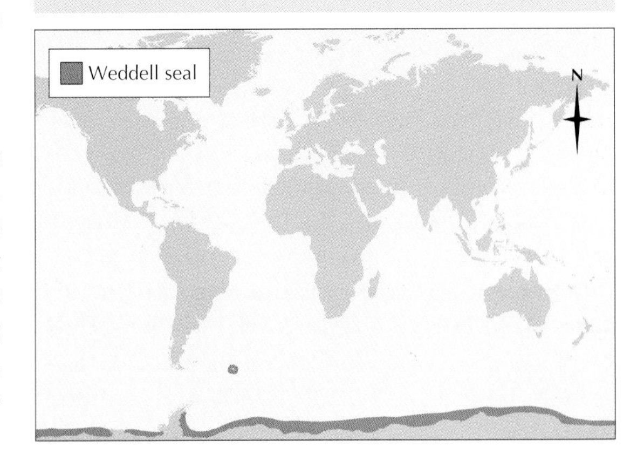

WEEVER

The lesser weever usually rests on the sea bottom, burying itself in the sand or gravel, with its eyes and the tip of the first dorsal fin exposed.

There are only eight species of the small weevers, or weeverfish, but these are so notorious they probably have more local common names than any other fish. The lesser weever's name, *Echiichthys vipera*, seems to be from the Latin *vipera* for a snake. Two other common names are stingfish and sea dragon.

The weever has a long body with a fairly small head, a wide mouth directed upward and large eyes that are well up on the head. The first dorsal fin is short and supported by a few spines. The second dorsal fin and the anal fin are long and low, and the tail fin is relatively large. The pelvic fins are small and lie forward of the largish pectorals. Each gill cover bears a stout, bladelike spine. The body is yellowish brown above, lighter on the sides and belly, with characteristic dark streaks on the sides that follow the oblique rows of scales. The first dorsal fin is mostly or completely black. The greater weever, *Trachinus draco*, is the largest at 18 inches (45 cm) in length, but most weevers are only a few inches long.

Weevers range from the coasts of Norway to West Africa and also into the Mediterranean, mainly from shallows to 300 feet (90 m) depth.

Night feeders

Weevers live on sandy bottoms, lying for much of their time buried in the sand with little more than the eyes and dorsal fin exposed. Weevers tend to be local in distribution, numerous in some places, sparse in others. Since they are often taken in nets at night they are probably more active at that time. In the Mediterranean the greater weever is numerous and is caught for food. In other places it may be used as fish meal.

Weevers themselves feed mainly on bottom-living animals, including crustaceans, such as shrimps and small crabs, and small fish, such as gobies, sand eels, dragonets and small flatfish. They also eat marine bristleworms and small bivalve mollusks. The greater weever's diet is composed mostly of fish, particularly sand eels.

Although the weever appears to be active mainly at night, it also feeds by day. Suberbly concealed in the substrate, it is able to pounce on passing prey, engulfing it with a snap of the jaws. Some scientists have suggested that the weever's bright eyes looking up from the sandy seabed lure fish down to the ambush.

Inflicting agony

The greater weever spawns from June to August; a slightly more extended season is found in the lesser weever. The eggs are ½₅ inch (1 mm) in diameter and float in the sea. They hatch in 5 to 10 days. The full life history of the weever is not yet fully known, perhaps understandably in view of the fish's venom equipment.

The weever is best known for the danger it presents to the unwary. The spines in the dorsal fin and those on each gill cover have venom glands at their bases. When lying buried, the fish erects the dorsal fin at the slightest disturbance, and bathers may easily tread on one. Each spine is sheathed in tissue with only its tip projecting, and there is a deep groove along the spine's margins. When the spine is stepped on, causing the tip to snap and tissues elsewhere to rupture, the venom floods into the wound. Shrimp fishers working inshore often take weevers in their nets and must be careful how they handle them. Even when weevers have died in the nets, their venom is still active, and an injury can cause excruciating pain. Envenomation is said to be fatal on occasion, but reliable records of this are hard to find. The venom is a nerve toxin that has distressing psychological side effects.

Only in defense

The fact that the weever can only inflict such pain when stepped on implies that the venom has a defensive, rather than offensive, purpose.

With its upward-tilting mouth and eyes on top of the head, the lesser weever is ideally adapted to a life on sandy, muddy or gravelly seabeds.

This idea is supported by observations showing that a weever does not use the venom on its prey. However, a fish attacking the weever is soon killed. For example, a goby that was observed attacking a weever died within 90 seconds, thrashing about violently and then turning belly-up. A blenny twice the size of a weever, attempting to swallow it, was seen to die within 2 minutes.

A warning to predators

Another reason for supposing the venom spines of weevers are purely defensive is the black color of the dorsal fin, which, when erected, stands out boldly against the yellows and pale browns of the weever's skin, or against the yellow sand when the weever is buried. The effect could be like the stripes of a yellowjacket's abdomen: a warning to predators not to touch. To be an effective defense against a predator a poison need not necessarily be lethal, but it must produce instant pain, taking effect before the predator's jaws have had time to do damage.

This aspect of weever biology was investigated in 1961 by Dr. D. B. Carlisle of the Plymouth Marine Laboratory, England. He collected the venom from the spines of a weever using a piece of sponge, and then injected small doses into his own arm. The pain was immediate and was followed by a rise in his pulse rate and respiratory distress. The symptoms were noticeable even after much diluted doses. Carlisle described the immediate pain as "more severe than that of any other venomous sting." The poison is due to 5-hydroxytriptamine, "one of the most potent of pain-producing substances."

WEEVERS

CLASS **Actinopterygii**

ORDER **Perciformes**

FAMILY **Trachinidae**

GENUS AND SPECIES **Lesser weever, *Echiichthys vipera*; spotted weever, *Trachinus araneus*; guinean weever, *T. armatus*; sailfin weever, *T. collignoni*; greater weever, *T. draco* (detailed below); striped weever, *T. lineolatus*; Cape Verde weever, *T. pellegrini*; starry weever, *T. radiatus***

WEIGHT
Up to 3½ lb. (1.6 kg)

LENGTH
18 in. (45 cm)

DISTINCTIVE FEATURES
Slender body; small head; yellowish brown upper body, paler sides and belly; short spines in front of and above each eye; 7 poisonous rays on first dorsal fin

DIET
Small invertebrates, small fish

BREEDING
Little known: egg-producing; pelagic egg and larval stages (occur in open sea)

LIFE SPAN
Not known

HABITAT
Sandy, gravelly or muddy seabeds to about 300 feet (90 m); also found in shallows of around 25–30 ft. (7.6–9 m)

DISTRIBUTION
Eastern Atlantic, from Norway south to Morocco, Madeira and Canary Islands; also Mediterranean and Black Seas

STATUS
Not threatened

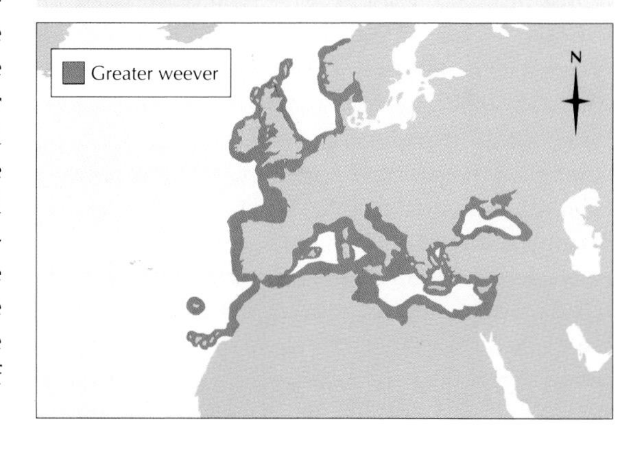

WEEVIL

INSECTS FORM THE MOST numerous and diverse class in the animal kingdom, and the beetles, which include the weevils, are the largest insect group. Entomologists have already described and named over 50,000 species of weevils and every year several hundred new ones are discovered; no doubt, there are many more still unknown to science.

Most weevils are only ⅛ inch (3 mm) or so long. A few North American species attain ½ inch (12 mm), and there are tropical weevils of about 3 inches (7.5 cm). They are generally compact, with the head drawn out into a snout or rostrum. In some species the rostrum is elongated, slender and down-curving, being especially long in the female; in others the rostrum is relatively short. The jaws are at the tip of the rostrum. Often the antennae arise from a point midway along the rostrum and are elbowed at the first joint so that they can be folded away into a groove on each side of the rostrum. Many weevils are wingless; of those that have wings, few fly much.

Egg-laying tube

All weevils, in both larval and adult form, are dependent on plants, either for food or as part of the life cycle. In the latter, a female uses her snout and jaws to drill a hole in a stem, bud or fruit. She then turns around and extends the long ovipositor (egg-laying tube) from the tip of her abdomen and deposits an egg deep in the tunnel. The larvae are tiny, white, legless grubs that live usually among the plant tissues. Some pupate within their larval habitat, whereas others gnaw their way out and become pupae in the soil.

Sometimes a female comes to grief as she bores holes for her eggs. Her feet may slip and she is then left poised on her embedded nose, her legs waving helplessly in the air. She has no means of getting herself out of this situation and remains there until she dies.

Old World weevils

Most weevils feed on some particular species or genus of plant, often on a certain part of the plant, and there are few plants without their associated weevils. This accounts to a great extent for the diversity of this family of beetles. The following are examples of weevils attached to particular plants.

The female nut weevil, *Curculio nucum*, of the Old World bores into hazelnuts while they are still green and lays an egg in each one. The larva feeds on the growing kernel until the nut drops in the fall, when the mature larva gnaws its way out and pupates in the soil. The proverbial bad nut is a hazelnut either containing, or vacated by, a little white, legless grub.

The figwort weevil, *Cionus scrophulariae*, is a small, brown beetle with a black dorsal spot. The larvae are unusual in feeding openly on the leaves. They are covered with slime and pupate while attached to the plant in cocoons formed of hardened slime. These cocoons resemble the seed capsules of the figwort and as a result probably derive protection against insectivorous birds.

The eggs of the gorse weevil, *Apion ulicis*, are laid in batches of a dozen or so in the seedpods of the gorse plant. The larvae feed on the seeds and pupate within the pods, hatching on dry days in summer when the gorse seed is ripe and the pods are splitting. In this plant the pods dehisce, or split, suddenly with a cracking sound and the seeds are scattered quite widely. Infested pods dehisce in just the same way, but in this case it is beetles instead of seeds that are hurled around. The weevils are thus spread as far as the gorse seeds on which they feed. This weevil has been introduced into New Zealand to control the rampant spread of introduced gorse.

One of the largest European weevils is the large pine weevil, *Hylobius abietis*, about ½ inch (12 mm) long, blackish, with patches of short, yellow hairs. It lives on conifers, especially pine, and the adult beetles damage saplings by feeding on the tender bark of the shoots and on the buds. The larvae are comparatively harmless, boring into the old stumps and roots of felled trees.

In some species of weevils, such as this nut weevil,* Curculio glandium, *the rostrum is as long as the insect's body.

A nut weevil,* Eupholus quinitaenia, *from Papua New Guinea. Each species is specially adapted to its own host plant.

WEEVILS

PHYLUM **Arthropoda**

CLASS **Insecta**

ORDER **Coleoptera**

FAMILY **Curculionidae**

GENUS AND SPECIES **Over 50,000 species, including: cotton boll weevil, *Anthonomus grandis*; gorse weevil, *Apion ulicis*; figwort weevil, *Cionus scrophulariae*; nut weevil, *Curculio nucum*; large pine weevil, *Hylobius abietis*; grain weevil, *Sitophilus granarius***

LENGTH
⅛–3 in. (0.3–7.5 cm)

DISTINCTIVE FEATURES
Beetles with head extended into elongate rostrum or beak with jaws at tip; strongly clubbed and sharply elbowed antennae part way along rostrum; often brightly colored

DIET
Roots, stems, leaves, seeds and fruits; some species eat rice and grain

BREEDING
Complete metamorphosis; eggs pupate in host plant or in nearby soil

LIFE SPAN
Usually a few weeks to 1 year

HABITAT
On or in host plants

DISTRIBUTION
Worldwide

STATUS
Common

Agricultural and food pests

The close attachment of particular weevils to particular plants has led to some of these beetles becoming agricultural pests. A weevil feeding on a wild plant that is brought into cultivation will almost certainly multiply enormously when its food plant, normally scattered among many other types of wild flowers, extends, uninterrupted, for hundreds of square miles. The pine weevil has become a pest in plantations of young pine, and in the southern United States the cotton boll weevil, *Anthonomus grandis*, costs the U.S. cotton industry between $1 million and $2 million a year.

This insect first invaded southern Texas from Mexico in 1892 and rapidly spread to all the cotton-growing regions. It is a typical weevil, compact, brown in color, ¼ inch (6 mm) long, and it has a stout, down-curved rostrum. Its eggs are laid in the buds or, later in the season, in the seed heads (bolls) of the cotton plant, one in each, and a single female may lay 100 to 300 eggs. The life cycle takes only 3 weeks, and in some localities there may be 10 generations in a year. The infested bolls cannot be used, and once inside the bud or boll, the larva is largely protected from sprayed or dusted pesticides. Carefully timed applications of insecticides are partly effective, and burning debris, in which the adults overwinter, is an important control measure.

However, the boll weevil can also be regarded as a benefactor of the southern United States. Before it invaded the area, agriculture was concentrated on cotton and tobacco, and so was very vulnerable to price fluctuations. The weevil forced the farmers to diversify, and the economy of the American South has benefited from this.

Another problem weevil is the grain weevil, *Sitophilus granarius*. The Anglo-Saxon word *wifel*, from which weevil is derived, referred to the grain weevil, which has infested stored grain since prehistoric times. The egg is laid, and the larva lives inside cereal grains of all kinds, hollowing them out and causing serious damage if uncontrolled. It can be destroyed by fumigation in suitably constructed storehouses. It also infests any foodstuffs prepared from flour. This weevil was a huge problem in the days of sailing ships, because the staple food on board was a thick, hard cookie in which the grain weevil thrived. Some sailors broke their cookies and tapped them on the table to dislodge the grubs and beetles; others ate them as they were and were possibly better nourished as a result.

WETLAND

Wetlands sustain a wide variety of wildlife. These scarlet and white ibises and black-bellied whistling ducks are flying over a flooded lake at the end of the wet season in the llanos *of Venezuela.*

Wetlands, occurring worldwide from the high Arctic to the most southerly subantarctic islands, are among the most biologically and geographically diverse of any biome. They are also among the most productive, some sites rivaling the tropical rain forests for their biomass (amount of living matter).

A wetland can be defined as an area where water (whether fresh, brackish or salt) is the primary factor controlling the environment and associated plant and animal life. A wetland is located where the water table is at or near the land's surface, or where the land is covered with water to a depth of a few yards. Thus, peat bogs and swamps, floodplains, salt marshes, deltas and mangroves are wetland biomes, but so are the riverine or coastal bodies of water bordering such wetlands, as well as coral reefs, lakes and rivers (covered in detail in separate guidepost articles). To the total can be added artificial wetlands, such as fish-farming pools, flooded gravel pits, rice paddies, reservoirs and canals.

Constantly evolving biome

Wetlands are in a constant process of change, influenced by both local and remote conditions. For instance, meltwater from mountains may be carried hundreds of miles along a river to wash over a level floodplain, depositing minerals and other nutrients that promote plant growth. Seasonal rains add to the flow, often causing widespread flooding over surrounding plains, where wetland vegetation, such as reeds, grasses and mosses, soaks up the inundation like a vast natural sponge, to release it in controlled amounts during the following dry season. The local climate regulates the rate at which water evaporates: rapidly in warm regions, more slowly at high latitudes or high altitudes. Coastal deltas, such as that of the Ganges River in the Bay of Bengal, may periodically be subjected to violent cyclones blowing in from the ocean, which can cause immense physical damage but also churn up and release nutritious sediments.

How much wetland?

Estimations of the extent of the world's wetlands vary considerably. The WCMC (World Conservation Monitoring Center), Cambridge, England, has suggested that wetlands cover 2¼ million square miles (5.7 million sq km), about 6 percent of Earth's land surface. Of this total, 2 percent represents lakes, 30 percent bogs, 26 percent fens, 20 percent swamps and 15 percent floodplains. It estimates there are some 93,500 square miles (240,000 sq km) of mangroves and about 234,000 square miles (600,000 sq km) of coral reefs. A global wetlands review of 1999, however, estimated a maximum of 17½ million square miles (45 million sq km) of wetland worldwide. The wide discrepancy between these two estimations is accounted for by a range of factors. The criteria by which an area is defined as a wetland may vary from country to country, or from year to year, as may the methods of data collection, resulting in very different figures. Based on the WCMC's estimation, Canada alone holds about 24 percent of the world's wetlands, totaling more than 500,000 square miles (1.3 million sq km), though these are spread out rather than focused in one major area. Alaska adds some 300,000 square miles (800,000 sq km) to the total, and the United States accounts for another 168,000 square miles (430,000 sq km). Though there are many kinds of wetlands in North America, those in the far north are dominated by bogs and fens overlying peat.

Peatland

Peat occurs on every continent, but is found in greatest quantities in the Northern Hemisphere within the latitudes 40–70° N. It is typically associated with swamps, floodplains, marshes, muskeg (mossy bogland found in northern North America) and fen. Peat is composed from layers of dead plant matter laid down over thousands of years. Low temperature and oxygen levels, low nutrient intake, high acidity and constant waterlogging combine to slow down or arrest the processes of decay by which organic matter is normally broken down and returned to the ground. Instead, the dead mosses and other plants simply build up, often to depths of 30 feet (9 m) or more. The cushionlike material has remarkable water-retaining properties.

Acid-loving plants, such as cranberry and sundew, thrive on peatland, which also supports swamp forests. The insect life of peat bogs is typically rich, attracting birds in high numbers. For example, the Teici Reserve in Latvia, eastern Europe, contains only 60 square miles (153 sq km) of bog, but it is home to 206 species of mosses and liverworts, 672 species of flowering plants and nearly 2,600 species of invertebrates. One-fifth of Teici is forested, mainly with Scots pine (*Pinus sylvestris*), silver birch (*Betula pendula*) and Norway spruce (*Picea abies*). It is home to nearly 200 bird species. Some of the year-round residents include the Arctic loon (*Gavia arctica*), lesser spotted eagle (*Aquila pomarina*) and black-tailed godwit (*Limosa limosa*). Those that visit to breed include the osprey (*Pandion haliaetus*) and wood sandpiper (*Tringa glareola*). Resident mammals include the gray wolf (*Canis lupus*), Eurasian lynx (*Lynx lynx*) and European beaver (*Castor fiber*).

Flooded plains

Away from urban areas, the regular flooding of level terrain adjoining major rivers is essential to wetland survival. This is the process that sustains the spectacular biological diversity of the Amazon River Basin, which covers 2¾ million square miles (7 million sq km) of tropical South America. North of Amazonia lie the llanos of Venezuela, a huge mosaic of sluggish watercourses, lagoons and marshes, bordering seasonally flooded grassy plains and palm savanna. From southern Brazil into Paraguay extends the Pantanal, which

***Exploitation of water resources in the Everglades has ruined the habitats of many species. A new $5.4-billion project aims to revive the area. Below: an American alligator,* Alligator mississippiensis *and great blue heron,* Ardea herodias.**

WETLAND

CLIMATE
Varies enormously, from subarctic and subantarctic to tropical; water is primary factor controlling environment

VEGETATION
Dominated by grasses, reeds, bulrushes and mosses; also mangrove trees along warm coastlines

DISTRIBUTION
Major wetlands include Everglades (Florida, North America); Amazon River Basin, Pantanal (South America); Okavango Delta (Botswana, southern Africa); Sundarbans (Bangladesh and India)

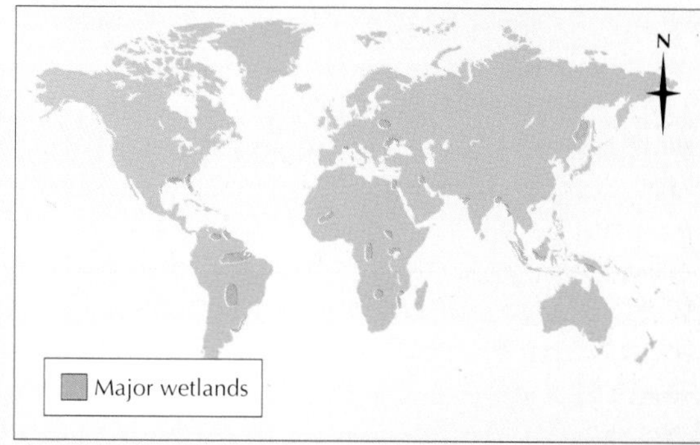

at 50,000 square miles (130,000 sq km) is the world's largest freshwater marsh. The wetlands fringe the banks of the Río Paraguay and many of its tributaries for hundreds of miles. The rivers' waters rise between December and May–June. By then the rainy season is over, and the *cerrado* (savanna) of the Pantanal has turned into marshland. The region is very important for more than 600 bird species, including the rare American finfoot or sungrebe, *Heliornis fulica*, and the roseate spoonbill, *Ajaya ajaja*, which has been hunted from parts of its original range.

One of North America's most important floodplains is the Everglades, at the southern tip of Florida. Most of its 2,260 square miles (5,850 sq km) lie within the Everglades National Park. Of its 1,000 species of seed-bearing plants several are tropical, including mangroves and palms. Here, the wetland is flooded not by a river, but by Lake Okeechobee. The summer rains herald the annual flooding, which spreads thinly over the Everglades, spreading nutrients and enabling insect, fish and crustacean populations to bloom. These in turn provide food for more than 300 bird species, including the bald eagle, *Haliaeetus leucocephalus*, a species becoming rare elsewhere in the United States but maintained in breeding sanctuaries in Florida. Resident Everglades mammals include the Florida panther, *Felis concolor coryi*, and manatee, *Trichechus manatus*, a large aquatic mammal. Both are currently regarded as under threat, along with other species including the American crocodile, (*Crocodylus acutus*), the snake bird (*Anhinga anhinga*) and the Atlantic Ridley turtle (*Lepidochelys kempii*).

In Africa, one of the best-preserved floodplain wetlands is the Okavango Delta in Botswana, southern Africa. Flowing southeast from Angola and the Caprivi Strip, the Okavango River disperses in the midst of the vast, arid Kalahari Desert. In the wet season, the delta may extend up to 8,500 square miles (22,000 sq km). Droughts are common, sometimes for several years in succession, but there are always enough permanent wetlands to sustain the dependent wildlife.

Mangroves and coral reefs

Mangrove trees are leathery-leaved species that live on brackish or saltwater coasts, proliferating at tropical and subtropical latitudes in the Americas, southern and Southeast Asia and Australasia. The warm, salty mud that collects around mangrove trees is low in oxygen, and the trees have adapted to various degrees in order to breathe. Some grow on stiltlike prop-roots that can perform gas exchange with the atmosphere; others extend bizarre, snorkel-like roots above the water. Mangroves are also, to varying degrees, salt tolerant. They have efficient filtration mechanisms to render as fresh as possible the water entering their roots and stems, and some species have salt-excreting glands in their leaves.

The largest unbroken mangrove forest in the world is found in the Sundarbans of Bangladesh and India, a sprawling labyrinth of water channels, islands and forests at the delta of the Ganges and Brahmaputra Rivers. Here, tigers (*Panthera tigris*) hunt chital or axis deer (*Cervus axis*) and wild boar (*Sus scrofa*), readily taking to the water. The gharial, *Gavialis gangeticus*, a slender-snouted crocodile adapted to hunting fish, lurks alongside its larger, deadlier relative the estuarine crocodile, *Crocodylus porosus*. The warm, protected waters that wash the mangrove roots along the Sundarbans' myriad creeks are the most important nurseries for the crustaceans caught off the entire eastern coastline of India. This story is echoed elsewhere, particularly in Southeast Asia, where fish (and rice, another wetland product) form the staple diet for millions of people. Mangroves are vital nurseries for fish fry (young). The forests are also a key source of fuel wood and roof-thatching materials.

Coral reefs are regarded as wetlands because they exist in shallow coastal waters, where the algae living with the coral polyps can receive the sunlight they need in order to photosynthesize. Reefs, like mangrove swamps, are ideal nurseries for marine animal life and like mangroves they act as breakwaters, protecting coastlines against the worst ravages of cyclones, hurricanes and tsunamis (super tidal waves). In Bangladesh, the government has replanted many mangroves in the Sundarbans to prevent disasters such as the 1985 storm surge that drowned 40,000 people.

Botswana's Okavango Delta supports many species of antelope (red lechwe,* Kobus leche, *pictured), as well as zebra, gazelles, ducks, geese, fish eagles and herons.

Many benefits of wetlands

Swamps, marshes and floodplains perform many services, often beneficial to humans. For instance, riverine wetlands are crucial for flood control. They hold heavy rainfall in the soil or in surface waters, preventing flooding downstream. Their capacity can be stretched to bursting point, however, by deforestation upstream.

Wetlands often also accumulate nutrients and sediment from river water. Valleys such as those of the rivers Indus, Tigris, Euphrates and Nile, enriched with sedimentary deposits, have been foci for human settlement from the earliest civilizations. Today, the sustainable harvest of wetland products can provide not only food but also fuel wood, construction materials, medical products and other staples. Wetlands also serve as gene pools for biodiversity, both for animals and for disease-free strains of crops such as rice.

Wetlands also help to recharge the underground aquifers that supply human populations with drinking water. Swamps, such as those in the Florida Everglades, as well as streams and rivers in rainfall catchment areas, perform this valuable task. Swamps and reedbeds are also highly effective filters, straining toxic bacteria from sewage or heavy metals from factory waste.

A less visible benefit of wetlands is the degree to which they absorb carbon, helping to mitigate the harmful effects of climate change.

Threats to wetlands

Worldwide, freshwater consumption rose sixfold in the 20th century. As more water is drawn from aquifers, the water table sinks and wetlands may dry out. The diversion of river water for human use is also damaging. The Sudd marshes in Sudan, northern Africa, escaped ecological disaster only when civil war in the 1980s halted construction of the Jonglei Canal, intended to divert water from the upper Nile for use in irrigation projects.

Pollution is a constant threat to wetlands. Acid rain and contamination from factory outflows kill the complex food chains in rivers, lakes and marshes. Agricultural fertilizers washed off the land cause algal blooms, choking life.

Peat is cut from bogs for use as a fuel, and peatlands are also used as grounds for conifer plantations. Mangroves, too, are exploited for fuel wood or timber, or converted to rice paddies. As they disappear, so do fish and crustaceans, the fry of which depend on mangroves as nurseries.

At the forefront of wetland conservation attempts is the Ramsar Convention on Wetlands of International Importance Especially as Waterfowl Habitat (usually known as the Ramsar Convention). This intergovernmental treaty, which came into force in 1975, has been signed by 188 parties. Its List of Wetlands of International Importance includes more than 1,000 sites covering some 285,000 square miles (730,000 sq km).

The Pantanal is the world's largest freshwater marsh. This photograph, taken in Mato Grosso, southwest Brazil, during the rainy season, shows swamp flora in bloom.

WHALEFISH

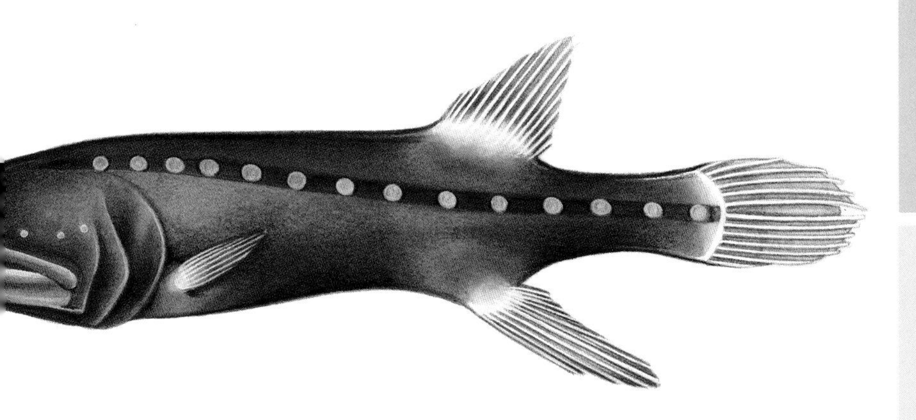

Below 3,000 feet (900 m) fish are blind or have degenerate eyes. It is therefore a mystery why whalefish, such as Cetomimus indagator *(above), have luminous patches.*

WHALEFISH ARE NOT named for their size but for their shape. They are small deep-sea fish, the largest being 9 inches (22.5 cm) long, and most are 4 inches (10 cm) or less. The head and relatively huge mouth also resemble those of the large whalebone whales. Whalefish are either blind or have very small degenerate eyes. Like so many deep-sea fish, whalefish are black, but they have brilliant patches of orange and red around the mouth and fins. The body tapers in the rear third to a relatively small tail fin. The dorsal fin is fairly large and soft rayed and so is the anal fin, which lies opposite on the underside of the tail. Both these fins have luminous patches believed to be due to a secretion from glandular patches at the bases of the fins. However their function is unknown. There are usually no pelvic fins and the pectoral fins are small.

Whalefish are placed in the order Cetomimiformes. The 23 species live at depths of about 6,600 feet (2,000 m) in tropical seas, from the Gulf of Mexico to West Africa and in the Indian Ocean to the western Pacific.

Detecting vibrations

To find their way about, whalefish have a highly sensitive lateral line made up of an enormous hollow tube communicating with the exterior by a series of large pores. They can probably detect the slightest vibrations in the water, which enables them to find their prey. Whalefish can swallow fish as large as themselves, capturing them by a rapid forward dart. They lack a swim bladder, and probably maintain a position in midwater by means of flotation appendages, typically cone shaped, which lie between the pores of the lateral line.

WHALEFISH

CLASS **Actinopterygii**

ORDER **Cetomimiformes**

FAMILY **Rondeletiidae, Cetomimidae, Barbourisiidae**

GENUS AND SPECIES **Redmouth whalefish, *Rondeletia loricata* (detailed below); velvet whalefish, *Barbourisia rufa*; others**

LENGTH
About 4¼ in. (11 cm)

DISTINCTIVE FEATURES
No spines above eye; usually orange and red coloration on black body; single dorsal and anal fins, set far back on body

DIET
Amphipod and crustacean remains

BREEDING
Egg-producing; eggs pelagic (found in open sea)

LIFE SPAN
Not known

HABITAT
Diurnal (daily) depths from about 6,600 ft. (2,000 m), migrating to 330 ft. (100 m) at night

DISTRIBUTION
Worldwide in tropical to temperate seas, of Atlantic, Indian and Pacific Oceans; also reported in waters around Iceland

STATUS
Not listed

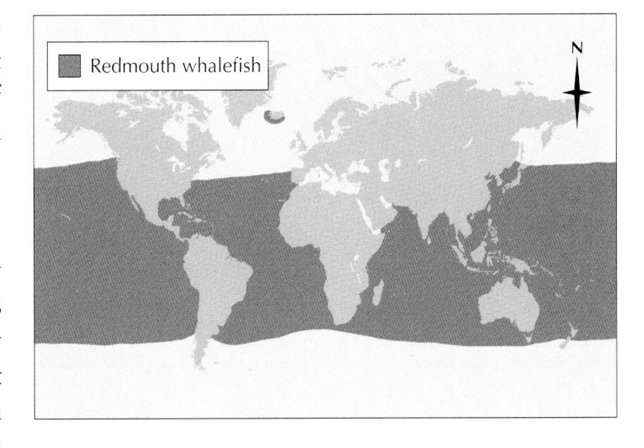

There is still much scientific research to be done regarding the life histories of whalefish, although scientists believe the larvae live in the surface layers of the ocean where light penetrates. Larval whalefish may have eyes that degenerate as the fish develop and sink to the ocean depths.

WHALES

Humpback whales are thought to suckle their calves for up to 1 year before weaning them.

WHALES ARE TRULY AQUATIC mammals and are found in the majority of the world's seas and oceans, from cold polar waters to the Tropics. Streamlined, powerful swimmers, they are ideally suited to their environment. Many whale species grow to an enormous size and include the largest creatures ever to have lived. For the last 300 years, humans have hunted whales, primarily for their fatty blubber, from which valuable oil may be extracted. This has led to the near extinction of several species, and subsequent worldwide protection for all whales.

Classification

Whales belong to the order Cetacea, which also contains the dolphins and porpoises. There are a total of 80 whale species within the order, and they are divided into two suborders: the Mysticeti or baleen whales, and the Odontoceti or toothed whales. Biologists believe that all whale species in existence today have evolved from coast-dwelling mammals that lived 50–60 million years ago.

This guidepost article concerns the three families of the suborder Mysticeti and four of the nine families of toothed whales in the suborder Odontoceti. Together, these animals are commonly thought of as whales. The remaining five families in the suborder Odontoceti comprise dolphins and porpoises, and are dealt with in a separate guidepost article. These are the true dolphins (Delphinidae), porpoises (Phocoenidae), river dolphins (Pontoporiidae), Indian river dolphins (Platanistoidae) and American river dolphins (Iniidae).

The suborder Mysticeti consists of three families of large, filter-feeding cetaceans. The family Balaenidae comprises four species: the northern right whale (*Eubalaena glacialis*), southern right whale (*E. australis*), bowhead whale (*Balaena mysticetus*) and pygmy right whale (*Caperea marginata*). Most authorities view the right whale populations in the Northern and Southern Hemispheres as distinct species. All four species of right whales seldom venture far from the coast, with the bowhead restricted to Arctic or subarctic waters. The family Eschrichtiidae contains only the gray whale, *Eschrichtius robustus*, which inhabits the coastal waters of the northern Pacific. The six species of rorqual whales, including the blue whale

CLASSIFICATION

ORDER
Cetacea

SUBORDER
Mysticeti: baleen whales;
Odontoceti: toothed whales

FAMILY
Balaenidae;
Eschrichtiidae;
Balaenopteridae;
Monodontidae;
Physeteridae;
Kogiidae;
Ziphiidae

NUMBER OF SPECIES
About 250

(*Balaenoptera musculus*) and humpback whale (*Megaptera novaeangliae*), belong to the family Balaenopteridae. Apart from Bryde's whale, *Balaenoptera edeni*, an inhabitant of the Tropics and subtropics, rorqual whales are found in all areas of all the oceans but rarely venture close to shore except to breed.

The sperm whale, *Physeter macrocephalus*, is the largest species within the suborder Odontoceti, and the sole representative of the family Physeteridae. Found worldwide, it shuns the colder seas and coastal waters. The family Kogiidae consists of the pygmy sperm whale (*Kogia breviceps*) and dwarf sperm whale (*K. simus*), which share their larger cousin's distribution. There are at least 20 species of beaked whales in the family Ziphiidae. Most of these live in the open sea and are little known. The family Monodontidae comprises two species of white whales, the narwhal (*Monodon monocerus*) and the beluga or white whale (*Delphinapterus leucas*), both of which are found in the Arctic. Many taxonomists also add the Irrawaddy dolphin, *Orcaella brevirostris*, an inhabitant of the coastal waters of southeastern Asia, to the white whale family.

Physical adaptations

Many whales are huge. The largest of all, the blue whale, reaches 72 feet (22 m) long and weighs up to 150 tons (137 tonnes). The largest toothed whale is the sperm whale. On average, male sperm whales grow to 60 feet (18 m) long and weigh 33 tons (29.7 tonnes). The advantage of being large is that the animal's surface area is proportionally small compared to body volume, so body temperature is stable and body heat is more easily conserved than is the case with smaller animals. On land, body size is restricted because an animal would eventually grow too heavy to move. Water, however, supports the body, so even the largest whale is practically weightless in its natural environment.

In order to live in the sea, whales have had to adapt the basic mammal form. A typical whale has a torpedo-shaped body designed to slip through water with minimum resistance. Its paddle-like forelimbs are used for steering, while its broad, flattened tail provides the power for swimming. A whale has no hair or fur, which would create drag in water and slow the animal down. The skin is made even smoother as the whale sheds oily skin cells, which act as a lubricant. With no hair to keep it warm, the animal relies on a layer of blubber for insulation.

In common with other mammals, whales breathe air, so they must return to the surface regularly to replenish oxygen. Whales' nostrils or blowholes are located on top of the head and are sealed when the animal submerges. Toothed whales possess a single blowhole, while baleen whales have two.

For most mammals, breathing air makes it impossible to remain underwater for any length of time. Whales, however, can spend up to 90 minutes submerged, at depths below 300 feet (100 m), where water pressure crushes the lungs. They achieve this by storing oxygen where it is most needed, in the muscles and blood, and allowing their lungs to collapse from the pressure.

Food and feeding

Toothed whales use their teeth to grasp large prey such as fish and squid, which are then swallowed whole. While baleen whales rarely dive more than 300 feet (100 m) to find swarms of their favorite krill or plankton, many toothed whales head into the ocean depths to find food. The narwhal and the beluga may dive to more than 500 feet (152 m) to find Arctic fish, while the sperm whale may descend to depths of 1¾ miles (3 km) in its quest for food. The larger the individual, the larger the prey item. Sperm whales have been known to devour rays, sharks and giant squid.

Visibility is poor at great depths. Consequently, deep divers rely on echolocation to find food. This process involves the animal emitting a rapid series of clicks, and listening for the returning echoes as the sounds bounce off objects around it. By assessing the way the echoes bounce back, the whale can build up a picture of its surroundings in order to navigate

Most great whales have barnacles, but the gray whale is especially prone to them. Whale lice live around the barnacles.

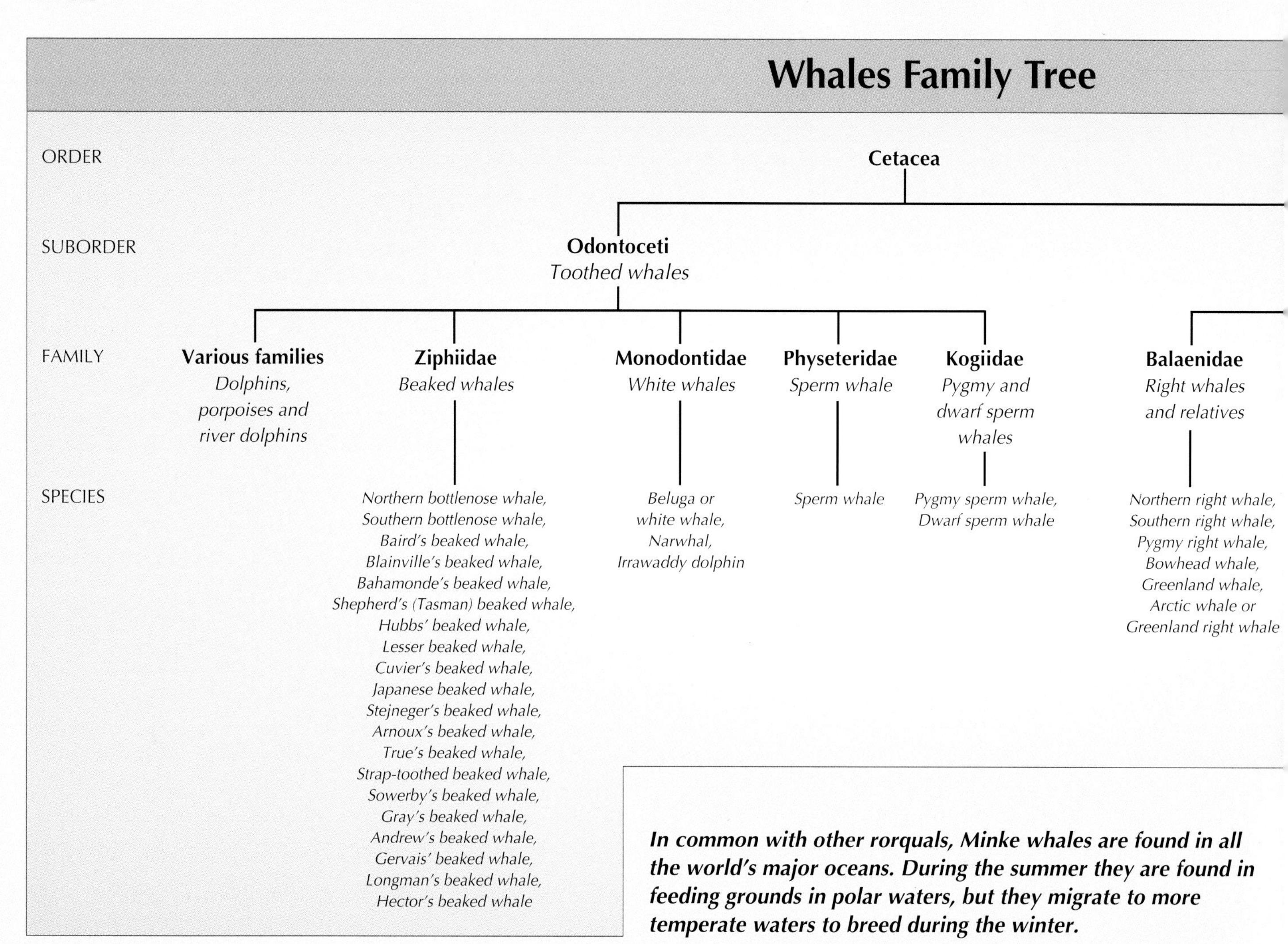

In common with other rorquals, Minke whales are found in all the world's major oceans. During the summer they are found in feeding grounds in polar waters, but they migrate to more temperate waters to breed during the winter.

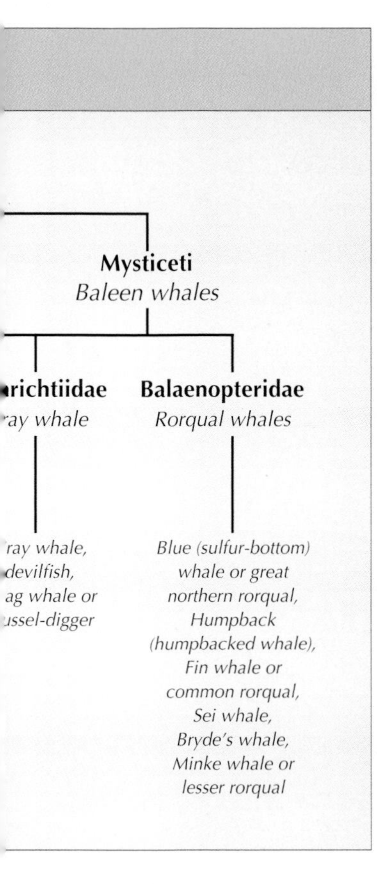

By assessing the way the echoes bounce back, the whale can build up a picture of its surroundings in order to navigate obstacles and locate potential prey.

A whale of the suborder Mysticeti uses baleen plates instead of teeth to snare prey. From close up, baleen resembles a dense mat of hair, and the whale uses this to strain food from the water. Made from keratin, a substance that is formed from hardened skin, the plates hang in the whale's mouth like curtains. Each species has a particular shape and length of baleen to suit its feeding technique. The right whale has long baleen plates. Its mouth is abnormally large in order to hold them. The whale skim-feeds by swimming with its mouth open just below the sea's surface. As the whale filters water through its mouth, the baleen sieves out krill and copepods (a type of small crustacean). The whale then scrapes the food from the baleen into its throat with its tongue.

A rorqual whale, such as the blue whale, gulp-feeds. Pleats in the tissue of the lower jaw region allow the whale to extend its mouth capacity and engulf an enormous mouthful of water. The whale then strains the water through its baleen, trapping any food. The gray whale has the smallest baleen plates. It feeds by scouring the sea bottom for mollusks and other small creatures.

Migration and breeding

Most whale species travel significant distances in search of food or suitable breeding areas, but it is only the baleen whales that follow strict migration routes. The gray whale, for example, makes an annual round trip of 7,500–12,500 miles (12,000–20,000 km) between Alaska or Siberia and the warmer waters off the coast of Mexico.

Baleen whales are forced to migrate due to the demands of feeding and breeding. The whales mostly feed in high latitudes, where the nutrient-rich waters sustain the vast swarms of krill and plankton that the large whales require. However, these polar regions are far too cold for the newborn whales, which have little or no insulating blubber. Therefore, females migrate to warmer waters in order to give birth and are usually accompanied by other females with young that have been born in previous years.

Whales barely eat during migration, surviving instead on fat reserves they have built up during the feeding season. From these reserves, females must not only feed themselves but also produce milk to feed their young. A female therefore loses up to 20 percent of her premigration body weight during the journey and breeding period. The super-rich milk ensures that the calf grows quickly and develops its own layer of blubber, which will enable it to survive the journey back to the feeding grounds.

Not all baleen whales migrate long distances. The bowhead whale stays in the Arctic all year, moving to the south of its range to give birth. Bryde's whale finds sufficient food in tropical waters and consequently makes no major migrations.

Because whale gestation takes 10–16 months, depending on species, males follow the females in order to mate so that the following year's young will be born at the correct time. They aim to breed with females who have recently weaned young and are in estrus (ready to breed). Battles between males for the right to mate can be fierce. Male narwhals, beaked whales and sperm whales have often been found with scars on their heads and body caused by the teeth of a rival.

Social behavior

All whales are social to some extent, and many travel in groups, called pods. Staying in contact in murky water or rough seas is vitally important and whales use a number of vocal and nonvocal means to communicate. Sperm whales, like most species, are tactile, and this probably helps pod

The humpback whale may breach (below) to communicate with other whales, to rid itself of parasites or to scare prey fish into concentrated groups that it may more easily attack.

Beluga whales usually form small herds of about a dozen but gather in their thousands around river mouths in summer.

members recognize and trust each other. Baleen whales tell each other who and where they are by producing a range of grunts, belches and squeals that carry for hundreds of miles. Toothed whales communicate with each other using clicks and whistles. The beluga's very audible chirps and whistles have led to it being called the sea canary.

Many whale species slap the water with parts of their bodies or leap from the water completely (breaching), possibly to communicate visually with other whales. Humpback whales tend to breach more often in stormy weather, when vocal communication is difficult. Some scientists argue that whales breach to rid themselves of parasites such as barnacles.

Whaling and whale survival

Few conservation issues generate as much emotion as whaling. Commercial whaling was banned in 1982, but the destruction wrought by whalers and the decision to resume whaling by several countries may still cause the extinction of many whale species. For hundreds of years whales have been a source of meat, oil and whalebone (baleen) for many people. Records show that Basque fishermen hunted large numbers of right whales in the Bay of Biscay from as early as the 10th century.

Until the 17th century, most whaling occurred in coastal regions and was not a global threat to whales. As Europeans explored more of the globe, however, whalers moved into the western Atlantic, using more sophisticated equipment to catch whales. By 1800, stocks of humpback, right and bowhead whales had collapsed in the North Atlantic and were soon to suffer the same fate in the North Pacific. Whalers moved on to the vast stocks of whales around the Antarctic.

The invention of the exploding harpoon and the steamship in about 1870 spelled doom for the rorquals, which until then had been too swift for the whalers. As the largest species, the blue whale was the first to be exploited, but when it began to decline whalers in turn hunted the fin whale, (*Balaenoptera physalus*), Sei whale (*Balaenoptera borealis*) and Minke whale (*Balaenoptera acutorostrata*). By 1910, whalers were using highly efficient factory ships to process whales at sea and so increase the number that could be harvested. Whaling reached its peak in 1937–1938, when 46,000 animals were taken in a single season.

Soon afterward, it became apparent that several species were approaching extinction and the International Whaling Commission (IWC) was set up to conserve stocks so that the whaling industry could survive. However, it achieved little until conservationists persuaded nonwhaling nations to join the IWC and vote for a total whaling ban. This was finally achieved in 1982, although Japan and Norway, among other nations, abstained. These countries maintained the right to take whales for what they claimed were scientific purposes. In 1993, Norway resumed commercial whaling.

Along with other threats, such as drift nets used by commercial fishers, increased ship traffic on migration routes and pollution, several whale species are still faced with an uncertain future. It may already be too late to save the blue whale, now reduced to a fragmented population of perhaps 6,000 to 14,000 individuals. Other species at severe risk of extinction include the right, bowhead, humpback, gray and fin whales.

For particular species see:
• BEAKED WHALE • BELUGA WHALE • BLUE WHALE
• GRAY WHALE • HUMPBACK WHALE • NARWHAL
• RIGHT WHALE • RORQUAL • SPERM WHALE

WHALE SHARK

The world's largest fish, yet harmless to humans, the whale shark can grow to a length of about 66 feet (20 m) and weigh up to 20 tons (20,300 kg). It is readily distinguished from any other fish by its prominent color pattern: very dark gray or brownish with white underparts, and the head and body covered in white or yellow spots that are smallest and densest on the head area. Rows of spots on the back are separated by pale vertical stripes. The whale shark has a long, cylindrical body with longitudinal ridges along its back, one down the middle and two or three on each side. Like all sharks, it has a very tough skin, that of a 50-foot (15-m) whale shark being 6 inches (15 cm) thick. The powerful tail is keeled and has an almost symmetrical fin. The head is broad and blunt, and the huge terminal mouth contains hundreds of very small teeth that form a type of rasp. The gape of the mouth is large, at about 5 feet (1.5 m) across in a medium-sized specimen. The eyes are small with small spiracles placed just behind them. The pectoral fins are large and sickle-shaped, and there are two dorsal fins, the second one lying above the anal fin.

The whale shark,* Rhincodon typus, *Ningaloo Reef, Australia. Whale sharks can be found singly or in groups of over 100 individuals.

One feature of the whale shark that is shared by only one other shark, the basking shark (discussed elsewhere), is the presence of gill rakers. The external gill openings above the base of the pectoral fins are very wide, and within the throat they are covered by close-set rows of sievelike gill rakers, each 4–5 inches (10–12.5 cm) long, growing out from the gill arches. They look like miniatures of the baleen plates of the whalebone whales and have the same function of straining out plankton and small fish.

Whale sharks are found in all the tropical waters of the world. Occasionally individuals have been reported as far north as New York and as far south as Brazil and in Australian waters.

Surface swimmer

The whale shark lives near the surface of the open sea, swimming slowly at around 2–3 knots. It is docile and does not attack swimmers, but it it is easily distressed by intruders, and the swish of its tail when fleeing is enough to stun a diver or break ribs. It also likes to bump against small boats, which are then at risk of capsizing. Whale sharks rub themselves deliberately against boats, possibly to get rid of external parasites.

Very few whale sharks have been caught, and they are not often seen except perhaps when basking at the surface. When wounded by a harpoon, the shark will dive straight down or streak away at speed, dragging the boat with it. It has very great powers of endurance and does not give in easily. It is said that if harpooned the whale shark can contract the muscles of its back to prevent the entrance of another spear. While swimming, the whale shark gives out a croaking sound, which is possibly a form of echolocation used in navigation.

Plankton feeder

The whale shark, like the basking shark, feeds on plankton, small schooling fish such as sardines and anchovies, and small crustaceans and squid. It does so by opening its huge mouth. Water rushes out over the gills, leaving the fish sticking to the inner walls of the throat and to the gill rakers. Although the whale shark feeds in this way, it still has numerous small teeth. They are arranged in some 310 rows in each jaw, but only about 10 or 15 rows function at any one time.

Stewart Springer of the U.S. Fishery Vessel *Oregon* once described seeing 30 or 40 whale sharks standing vertically, head up and tail down, during a spell of calm weather in the Gulf of Mexico. They were pumping up and down in the water, feeding on small fish and accompanied by small black-fin tuna that had stirred up the sea all around with their darting and leaping. When actively feeding on zooplankton the whale sharks turn their heads from side to side with part of the head lifted out of the water, opening and closing the mouth and gill slits.

When feeding on plankton the whale shark turns its head from side to side, opening and closing its mouth up to 28 times a minute. The current of swallowed water causes the gill covers to beat in time.

Live young

Whale sharks are ovoviviparous, producing eggs that, when developed, hatch within the mother's body or immediately after emerging. Late-term embryos shed their egg case inside the uterus at a size of 23–25 inches (58-64 cm). The smallest free-swimming young are 21½–22 inches (55–56 cm) long and have umbilical scars. A pregnant female was recently found with 300 embryos inside her, the largest of which measured 23–25 inches (58–64 cm) long.

Few predators

The whale shark has few natural predators. Because of its large size, only the sea's largest carnivores would attempt to attack it, and a blow from its powerful tail would probably be enough to drive away even the largest predator. The whale shark is not hunted commercially by humans; even its liver oil, the prize for many shark fisheries, does not contain vitamin A.

Gentle giant

The whale shark could truly be called a giant that does not know its own strength, and this is especially illustrated by the following. In 1919 a whale shark became wedged in a bamboo stake-trap set in water 50 feet (15 m) deep in the Gulf of Thailand. It appeared to have made no attempt to break its way out. In the same area in 1950 another whale shark was captured and beached by the local fishermen, and although details of its capture are not at hand, it would seem that the giant fish offered little or no resistance. Prince Chumbhot, reporting this incident, says it was "towed out to deep water and released by fishermen as a matter of luck, with a piece of red rag tied round its tail."

WHALE SHARK

CLASS **Elasmobranchii**

ORDER **Orectolobiformes**

FAMILY **Rhincodontidae**

GENUS AND SPECIES ***Rhincodon typus***

WEIGHT
Maximum 20 tons (20,300 kg)

LENGTH
Up to 66 ft. (20 m)

DISTINCTIVE FEATURES
Huge, blunt-headed shark; terminal mouth; prominent pattern of white or yellow spots and stripes on dark background; spots smaller and closer together on head; gill rakers; numerous rows of small teeth

DIET
Planktonic and swimming prey: small crustaceans, fish and squid

BREEDING
Few details known; ovoviviparous, giving birth to hatched or soon-to-hatch young; embryos up to 23–25 in. (58–64 cm) long

LIFE SPAN
Not known

HABITAT
Usually offshore surface seas, sometimes close inshore in lagoons or coral atolls

DISTRIBUTION
Tropical waters: western Atlantic, from New York through the Caribbean to central Brazil; eastern Atlantic, from Senegal to Gulf of Guinea; Indian Ocean; western Pacific, from Japan to Australia and Hawaii; eastern Pacific, from California to Chile

STATUS
Not listed as threatened

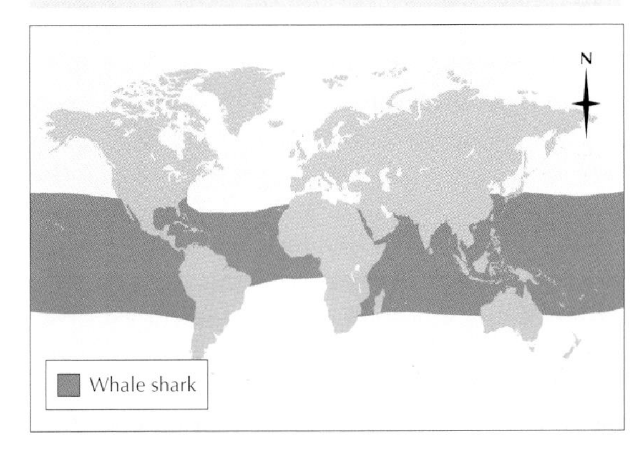

WHELK

THE NAME WHELK WAS originally used for one particular sea snail, the common or white whelk, *Buccinum undatum*, of Europe, also called buccinum in North America, where it occurs along the North Atlantic coast from the Arctic to New Jersey. In Scotland it is called the buckie. The name is derived from the Latin *buccina*, meaning a twisted trumpet. It was originally applied to almost any trumpet-shaped shell, and the name whelk is still sometimes loosely used in this way.

The common whelk has a thick, chalky gray to yellowish shell up to 4¼ inches (11 cm) long, and exceptionally 6 inches (15 cm) long. It is just one member of the family Buccinidae, which contains up to 50 genera and several hundred species. The common whelk shell figured in many textbooks of the 19th century because of the way it is used as a mode of transport by sea anemones. Another European species is the red or almond whelk, *Neptunea antiqua*. Its shell is usually 4½ inches (11.3 cm) long, but it may be as long as 8 inches (20 cm). Another red whelk, *Charonia rubicunda*, is unrelated to true whelks and lives in Australian seas.

One of the best-known whelks in the United States is the channeled whelk, *Busycon canaliculatum*, found from Cape Cod to Florida. Its shell is 5–7½ inches (12.5–19 cm) long. Forming a separate family, Nassariidae, are the dog-whelks or mud snails, often attracted to lobster pots in the course of their scavenging. These dog-whelks are distinct from the European dog-whelk, *Nucella lapillus*, which belongs to the family Muricidae. Its shell, 1–1½ inches (2.5–3.8 cm) long, may be yellow, white, mauve or brown and is often banded. The members of the family Muricidae bore holes through the shells of other mollusks and then insert their long proboscises and eat the flesh, or else they pry open bivalve shells using a tooth on the lip of their own shell aperture. The American whelk-tingle, rough whelk or oyster drill, *Urosalpinx cinerea*, is another of the same family and a serious pest of oysters. It was introduced into British waters with American oysters in 1920. Whelks of this family produce a dye that was once highly valued, Tyrian purple. Other whelks include the heavy whelks of the Vasidae family of Australia and the distantly related needle whelks of the family Cerithiidae.

Not suited to shore life

The common whelk is found on every type of sea bottom from near low water to great depths, and its abundance has in the past made it important as food. Although large empty whelk shells are familiar objects on the shore, the larger living individuals normally live offshore and only the small ones occur between tidemarks, usually in well-sheltered places where there is mud between large stones. The common whelk is not suited to life on shore. When uncovered by the tide, it may cling to a rock until the tide returns. On mud or sand, however, it continues to crawl about, perhaps extending its foot and waving it around instead of retreating into its shell and using the operculum to seal the aperture as would a winkle. Therefore the water drains from the gill cavity and the tissues gradually dry out. In the Bay of Fundy, on the eastern coast of Canada, where the tidal range is the greatest in the world, many of these whelks are left exposed on the shore during the great ebb of the spring (extreme) tides and consequently die.

The striped dog whelk,* Nassarius glans, *of Papua New Guinea. Whelks are preyed on by bottom-feeding fish, such as rays and dogfish.

Long feeding tube

The common whelk feeds by means of a long proboscis, which may be extended for about 2 inches (5 cm) or even twice the length of the shell. At its tip is the mouth with its rasplike radula. The whelk's food includes fresh carrion, crabs, worms, bivalve mollusks and fish. It finds these with a long siphon that waves about as it samples the taste of the water. The siphon also brings clean water into the gill cavity while the whelk is buried or feeding on rotten flesh. Plaice caught in nets are sometimes attacked by whelks;

The common whelk inhabits a range of hard and soft substrates but is most abundant on mixed seabeds. It can also penetrate estuaries to some degree.

the whelk inserts its proboscis directly into the flesh of the fish, and sometimes 10 to 20 whelks may attack a single fish.

Whelks also insert the outer lips of their shells between the valves of cockles, scallops, mussels and oysters to prevent them from closing and to enable the radula to rasp away inside. The channeled whelk also feeds on bivalves, but if the valves cannot be wedged apart, the whelk chips away at their edges with its own shell until it can force in the proboscis.

Young whelks feed on eggs

As familiar as whelks themselves are the masses of usually empty egg cases that are so often cast ashore. Sometimes called sea wash balls, they are said to have been used by sailors as sponges, which they closely resemble. Before their true nature was realized, they were named as a species of coralline, *Alcyonium*.

Breeding occurs in late fall, and egg masses are laid soon after, hatching in February and March. The female whelk turns herself around as she deposits the eggs in a mass of capsules. Although the lower egg cases in the mass are usually fastened to some hard object, the mass often breaks free, and many empty egg cases are eventually washed up on the shore.

A single female produces up to 3,000 capsules, each about 1 inch (25 mm) across and joined to its neighbor by projections around the edge. However, several females may together produce a mass of capsules over 1 foot (30 cm) across, made up of as many as 15,000 capsules. Smaller-sized capsules are usually produced by smaller whelks.

Each egg capsule contains several hundred eggs, occasionally more than 3,000, each about 0.3 millimeters across. From these capsules only 10 to 15 young snails emerge. This is because the remaining eggs, although fertilized, serve to nourish those few embryos that are destined to become small whelks. These eggs are called nurse cells or food eggs. The snails that finally emerge after about 2 months are about ⅛ inch (3 mm) long. However, some of the unhealthy, developing young whelks may be devoured by their healthier companions before hatching.

COMMON WHELK

PHYLUM **Mollusca**

CLASS **Gastropoda**

ORDER **Neogastropoda**

FAMILY **Buccinidae**

GENUS AND SPECIES ***Buccinum undatum***

ALTERNATIVE NAMES
Whelk, buckie, northern buccinum, common northern buccinum

LENGTH
Usually up to 4¼ in. (11 cm)

DISTINCTIVE FEATURES
Robust, heavily coiled, fairly pointed shell with 7 or 8 whorls; rough shell surface

DIET
Crabs, bivalves and fish; carrion

BREEDING
Age at first breeding: 2–3 years; breeding season: late fall; number of eggs: up to 3,000, hatching February–March

LIFE SPAN
Probably more than 10 years

HABITAT
Various seabed substrates, such as stones or shells over sand and mud, from low water to depths of 4,000 ft. (1,200 m)

DISTRIBUTION
Cold waters of the North Atlantic

STATUS
Common

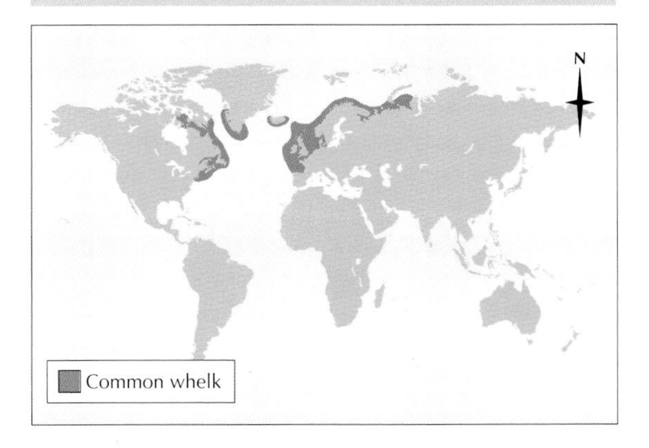

WHIP SNAKE

THE NAME whip snake, usually spelled whipsnake in the United States, denotes a group of long, thin, active ground-dwelling snakes. It is used mostly to describe such snakes in North America, Europe, parts of Asia and Australia. They are not necessarily related to one another; most whip snakes belong to the family Colubridae, but Australian whip snakes belong to the family Elapidae.

The three North American species are very closely related to racers (discussed elsewhere in this encyclopedia). Coachwhips, *Masticophis flagellum*, have the widest distribution. They are found from South Carolina and Florida westward to California and also in much of Mexico. The largest recorded coachwhip was more than 8 feet (2.4 m) in length, but most individuals are only half this size. They are very variable in color and pattern. In the eastern part of their range, many coachwhips are dark brown or black at the front end but light brown at the rear. This feature is unique among North American snakes. Striped whip snakes, *M. taeniatus*, are slightly smaller than coachwhips. They too are highly variable in appearance, but a pale stripe running along each side of the body is characteristic, although not seen in all individuals. Their distribution extends along a band from Washington State through Texas into the eastern half of Mexico. Sonora whip snakes, *M. bilineatus*, are found only in Arizona and the adjacent part of Mexico.

Europe has six species of whip snakes. The best-known species is the dark green whip snake, *Coluber viridiflavus*, which on the mainland is found in France and Italy, and also on the islands of Sicily, Sardinia and Corsica. Many individuals are dark or olive green in color, with speckled yellow markings.

Eight species of whip snakes are found in Australia, although there are a number of closely related species that have no English names. They all belong to the family Elapidae, and are venomous. Most of these snakes occur only in the northern parts of Australia, but the yellow-faced whip snake, *Demansia psammophis*, is found in every state except Tasmania. Yellow-faced whip snakes are rarely more than 2½ feet (75 cm) in length. They too are variable in appearance. Many individuals are pale, olive or yellow gray in background color, but each scale is lined with black, giving a reticulated appearance. There is usually a black stripe that runs from the eye to the angle of the jaw. This has a yellow band on either side of it, and it is these bands that give the snake its name. Yellow-faced whip snakes feed almost exclusively on lizards.

The largest dark green whip snakes may reach more than 6 feet (1.8 m) in length, but most adults are only two-thirds this size.

Difficult to capture

A common characteristic of the whip snakes on all continents is that they can move very fast. This enables them to catch the fast-moving lizards on which many of them feed. The prey are not constricted first, but are swallowed immediately. Whip snakes can disappear into undergrowth or under stones with remarkable agility and this can make them very difficult to capture. Lataste, a 19th-century French authority on snakes, became familiar with the habitual basking place of a dark green whip snake in France. However, although he saw this snake many times over a period of two years, he could never catch it. The Lataste's viper of the Iberian peninsula is named after this man.

Curious distribution

Coachwhips are found in most of the southern half of the United Sates but, curiously, do not occur along a corridor that runs on either side of the Mississippi River. One would expect the appearance of the highly variable snakes to differ on either side of this break in their distribution, but it does not. The reasons for this corridor are not known. Boundaries can, however, be found in the distribution of different forms: one lies about 400 miles (650 km) west of the Mississippi, running through Kansas, Oklahoma and eastern Texas. Snakes on one side of this line are often

Unlike many biting snakes, whip snakes do not puncture the skin then let go. They keep the mouth closed, yank the head, and lacerate the skin. The saliva may cause a slight swelling.

called eastern coachwhips, while those to the other side are called western coachwhips. The differences between them are mostly technical and are related to details of the arrangement of the scales, but they are distinct enough for some biologists to regard them as separate subspecies. Coachwhips in California, Nevada and Arizona also differ sufficiently for them to be described as a separate subspecies. These are often reddish brown in color and, confusingly, are sometimes called red racers. As with rat snakes (discussed elsewhere), coachwhips are so variable that many subspecies have been described, but there is much disagreement about how they should be named. The argument is not yet settled, and the names may well change again in the future.

Saint Paul's viper

The North American and European whip snakes are very aggressive, as are some of the species in Asia. The green whip snake of India and Sri Lanka is widely believed to strike at the eyes, and its Singhalese name means "the eye-plucker." The name is well earned, for although whip snakes do not necessarily aim for the eyes, they strike aggressively with the mouth open, even attempting to strike through the glass of a cage. Unlike many other aggressive snakes, whip snakes do no become tame in captivity. If it is cornered, a European dark green whip snake hisses and lunges at its attacker repeatedly and bites if it gets the opportunity. However, the snakes are not venomous.

Whip snakes are abundant on a number of Mediterranean islands, including Malta. It has been suggested that a whip snake was the viper that bit and clung to Saint Paul's hand when he was shipwrecked there. According to the story in the Acts of the Apostles, local people were surprised that Saint Paul came to no harm from the bite and regarded him as a god. In fact, true vipers, which are venomous, do not occur on Malta.

WHIP SNAKES

CLASS **Reptilia**

ORDER **Squamata**

SUBORDER **Serpentes**

FAMILY (1) **Elapidae (Australia)**

FAMILY (2) **Colubridae (elsewhere)**

GENUS AND SPECIES **North America: Sonora whip snake, *Masticophis bilineatus*; coachwhip, *M. flagellum*; striped whip snake, *M. taeniatus*. Europe: dark green whip snake, *Coluber viridiflavus*. Asia: green whip snake, *Ahaetulla mycterizans*. Australia: yellow-faced whip snake, *Demansia psammophis*. Others.**

ALTERNATIVE NAMES
Prairie runners, red racers (coachwhips); ornate whip snakes, cedar racers (striped whip snakes); western whip snakes, yellow-and-green snakes (dark green whip snakes)

LENGTH
Coachwhip may grow to 8 feet (2.4 m); most adults half this size

DISTINCTIVE FEATURES
Long, thin, fast-moving snakes; many species aggressive; only Australian species venomous

DIET
Lizards; small mammals and birds

BREEDING
Oviparous (lay eggs), except some Asian species

LIFE SPAN
Not known

HABITAT
Diverse; usually avoid dense forest and jungle

DISTRIBUTION
North, Central and parts of South America; central and southern Europe; Asia; Australia

STATUS
Abundant in many places

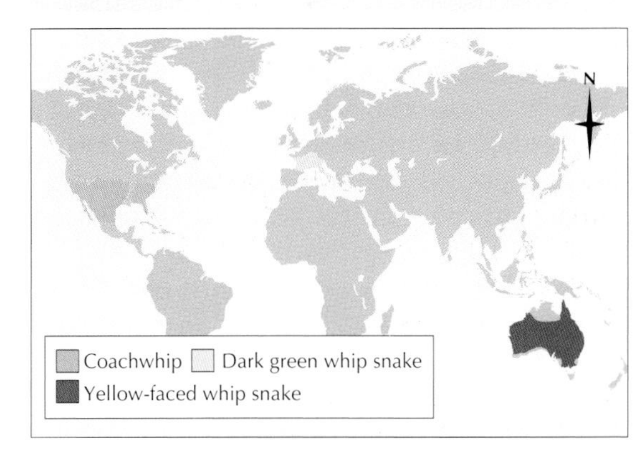

WHIRLIGIG BEETLE

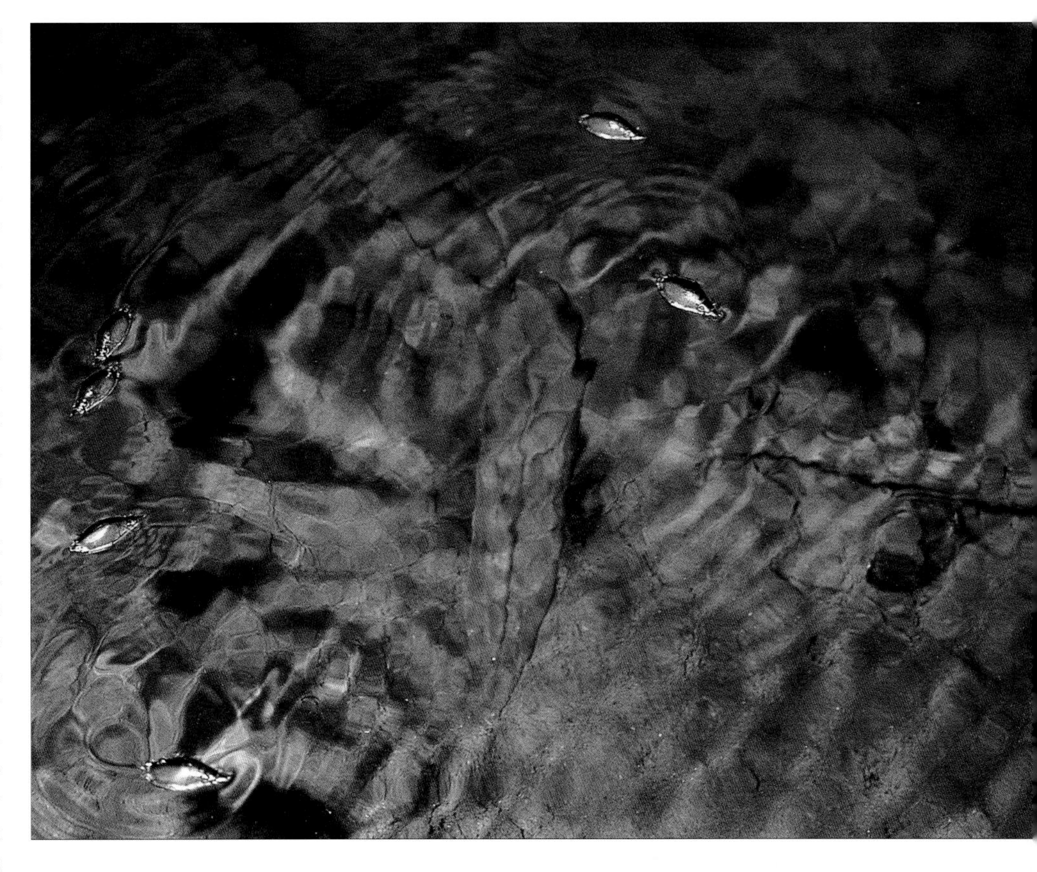

THE SMALL BEETLES THAT gyrate rapidly on the surface of the water in ponds and canals are called whirligig beetles. The name dates from the 15th century: originally, a whirligig was a child's toy that spun. Most of the whirligig beetles belong to the genus *Gyrinus* and are shiny bluish black or dark bronze in color. They swim only with the second and third pairs of legs, which have their segments flattened and fringed with hairs, making them effective oars. The first pair of legs is slender and not adapted in this way. This pair plays little or no part in swimming.

Whirligig beetle antennae are short and clubbed. The eyes are notable in that they are divided into separate lower (ventral) and upper (dorsal) compound eyes on each side of the head, and scientists assume that the facets, or ommatidia, are adapted for aquatic and aerial vision respectively. This arrangement effectively gives the beetles four eyes: two to watch what is going on in the water below them and two to observe what is going on in the open air. Whirligig beetles have wings and fly readily. Indeed, they are likely to quickly escape if placed in an open aquarium.

The family Gyrinidae, containing a total of about 400 species, is a relatively small beetle family compared, for example, with the Curculionidae, which contains the 50,000 species of weevils. Although most whirligig beetles are shiny, in the genus *Orectochilus* the insects' upper surface is covered with thick, yellowish gray hair. *Orectochilus* species are known as hairy whirligigs and live in running water, gyrating on the surface only at night. By day they hide under banks. A number of hairy whirligig species are found in Asia, while *O. villosus* is the only species of the genus found in Europe, and only one species is known to occur in Africa.

Semisubmerged skaters

Their capacity for scooting across the water's surface, weaving around in dense crowds but never colliding, is the whirligigs' most characteristic feature. They do not travel over the surface film as do pond skaters, family Gerridae. Although the shiny or hairy upper surface repels the water and remains dry, the whirligig beetles' underside and legs are immersed in the water. The antennae, meanwhile, are held in the surface film and are believed to be sensitive to changes in its curvature, thus enabling the beetles to avoid collision with each other. Each beetle makes a dimple in the film around it and throws up little bow waves in front. The two pairs of swimming legs do not rely only on their fringe of hairs for propulsion, as do those of most other water insects. The flattened joints fold up like a fan on the forward movement and then open up for the backward propulsive stroke.

Maintaining position

Observation of a group of whirligigs shows that the insects can maintain their position in a pond, or even in slowly flowing water, without simply scattering, as would perhaps be expected from each individual's mazelike movements. There seems to be only one possible means by which the beetles can hold their position when swimming. They must keep a constant pattern of objects on the bank in view. The fact that they can do this while weaving about in elaborate convolutions implies a remarkable degree of coordination of movement and vision.

Whirligig beetles can swim well underwater and dive readily when they are alarmed, carrying a bubble of air attached to the hind end.

Whirligig beetles are holometabolous insects: they go through a complete metamorphosis as they pass from the larval stage to adulthood.

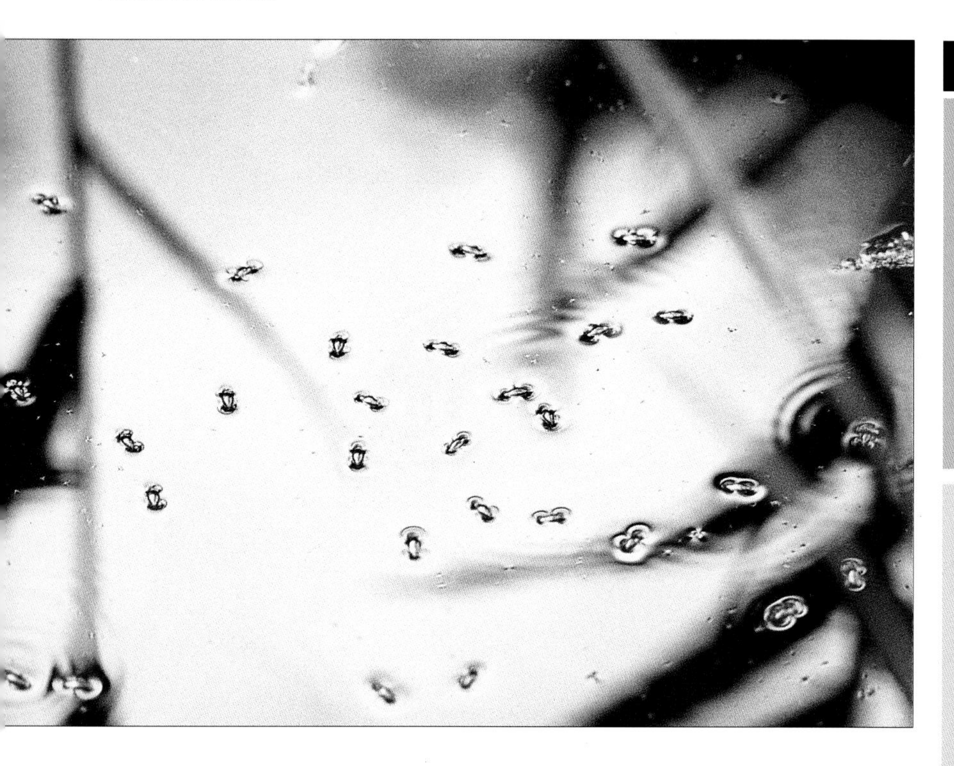

Whirligig beetles of the genus Gyrinus *scoot about the water surface. Of the 400 species of whirligigs, about 50 from 3 genera, including* Gyrinus*, occur in North America.*

WHIRLIGIG BEETLES

PHYLUM **Arthropoda**

CLASS **Insecta**

ORDER **Coleoptera**

FAMILY **Gyrinidae**

GENUS **Several, including *Gyrinus, Orectochilus***

SPECIES **About 400, including *Gyrinus natator; Orectochilus villosus***

LENGTH
1/10–6/10 in. (3–15 mm)

DISTINCTIVE FEATURES
Small, black, usually shiny insect; middle and back legs flattened and covered in hair; characteristic skimming motion on water

DIET
Small insects falling onto water's surface; also underwater insects

BREEDING
Eggs laid on submerged water plants; larval stages underwater; pupation in cocoons on plant stems above water surface; adults emerge about 1 month later

LIFE SPAN
Usually 1 year

HABITAT
Ponds; canals; slow-moving water

DISTRIBUTION
Worldwide

STATUS
Common

This bubble is probably used for respiration when the beetles rest, clinging to underwater plants. Whirligigs make no attempt to hide themselves at any time and, for all their active movement, seem to be easy prey for birds and fish. However, they probably are afforded some protection by their ability to exude a milky fluid when molested. This fluid is variously described as smelling disagreeable or, in one American whirligig, *Dineutes americanus*, like apples, but it is likely to function as a repellent in all cases. So far as is known, whirligigs themselves feed on small insects that fall onto the water surface, which they seize with their long, mobile front legs. They occasionally dive after insects, too.

Aquatic larvae

Whirligigs overwinter as adults, and the females lay their eggs in spring, end to end in rows on submerged water plants. The larvae hatch and grow rapidly. They swim in a wriggling manner or creep about the bottom, feeding on other insects and to some extent on vegetable matter. Their pointed jaws are hollow for sucking the juices of their prey. The larvae of hairy whirligigs are found among gravel in shallow flowing water and, like the adult, keep out of sight.

When fully grown, a whirligig beetle larva is very slender and more than ½ inch (13 mm) long. The joints between the segments of the larva are deeply indented and there is a row of feathery gills along each side of the body, one pair on each abdominal segment except the last, which has two pairs. The last segment also has two pairs of hooks, which scientists believe are used for climbing about on water plants. Close to the end of July, the larvae climb up the stems of emergent water plants and spin cocoons above the surface. The adults appear a month or so later, and it is at this time, in late summer, that whirligigs are extremely abundant.

Aquatic all-rounder

A whirligig beetle is unusual in enjoying the best of three worlds. Although it spends much of its time sculling about on the surface, it can easily take off and fly, skimming over the surface to look for a fresh feeding ground. To land, it noses over into a steep dive, using its wing cases as a parachute, and alights gently on the water. It can then, if necessary, shut its wing cases, trapping a bubble of air under them, and submerge like a submarine or a scuba diver.

WHISTLER

THE WHISTLERS, OR THICKHEADS, are small birds with melodious, often whistling songs. They are related to the flycatchers. and are found in Australia, New Guinea, Malaysia, the Philippines and the islands of the South Pacific. There are 57 species of whistlers, including six shrike-thrushes, six pitohuis, the wattled ploughbill, *Eulacestoma nigropectus*, and the morningbird, *Colluricincla tenebrosa*.

In Australia, some whistlers have alternative names such as robin, shrike-thrush or tit-shrike. As the alternative names suggest, they have large heads and rather shrikelike bills. Some have crests and a few have wattles or bare patches of skin. The whistlers vary in size from 6–13 inches (15–32.5 cm). The plumage usually contains yellow or green and black and the throat is usually white.

In common with most other whistlers, the golden whistler, *Pachycephala pectoralis*, is sexually dimorphic (each sex has distinctive features). The male has a bright plumage that contrasts with that of the dull female. The male has a white throat, which is separated from the yellow breast and abdomen by a narrow black band that joins the black on the head. The back is bright olive green, and the wing tips and tail are gray and black. The female has gray-brown upperparts and gray-white underparts.

The golden whistler ranges from Peninsular Malaysia south to Australia and the Fiji Islands. The rufous whistler, *P. rufiventris*, is another Australian whistler. The male is chestnut whereas the golden whistler is yellow. The black band extending across its chest continues along each side of the throat to the bill. The female is streaked with white on the abdomen.

Birds of ground cover

Whistlers mainly inhabit forest and scrub. The golden whistler is found in open green forests and in wattle scrub, while the white-breasted whistler, *Pachycephala lanioides*, of Australia, is found in the mangrove swamps of the northern coast. Whistlers generally choose to live in areas where there is good cover, and spend their time on or near the ground. Outside the breeding season they gather in flocks, often with other birds. Whistlers are usually shy birds, and ornithologists still have much to learn about their habits. The golden whistler, however, is better known than many other whistlers because it frequently enters suburban gardens and may become quite tame.

Mainly insect eaters

Most whistlers feed on insects, such as beetles and grasshoppers, which they collect among foliage or on the ground. The whistling shrike-thrush, *Colluricincla rectirostris*, peels bark with its bill to find insects and their larvae underneath. A few whistlers eat berries, and some of these lack the rictal bristles around the base of the bill, which are generally an adaptation for sweeping up insects and probably are a disadvantage in berry-eating.

Varied songs

The songs of the whistlers are among the finest of any Australian bird. Early European settlers called the golden and rufous whistlers thunderbirds because loud noises made them sing. The rufous whistler sometimes sings to defend its territory year-round. It has a number of phrases in its song, including sounds similar to a whip-crack. Other whistlers, such as the olive whistler, *P. olivacea*, have wistful calls. The golden whistler has a call that resembles someone whistling to a

The bare-throated whistler, **Pachycephala nudigula,** ***is endemic to Sumbawa and Flores, Indonesia.***

As with other whistlers, the bare-throated whistler lives mainly in forests and scrub, up to about 5,000 feet (1,500 m).

dog, the notes usually being represented as *wi-wi-wi-wi-wit*. Male whistlers defend their territories by singing and performing bowing displays, and the female rufous whistler assists in the defense. In this species, both sexes sing during courtship, the female answering the male's whip-crack calls.

The cup- or saucer-shaped nest is built of bark, dead leaves and dry grass, and is lined with feathers or fibers. It is usually placed in an upright fork of a tree. There are two or three white or brownish eggs, which are incubated by both parents or by the female alone for about 15 days. Both parents care for the young, which stay in the nest for two weeks, and sometimes perform distraction displays to lure away predators.

Record varieties

There are about 64 subspecies of golden whistlers, a record among birds, probably because its range from Java to the Fijis includes a vast number of islands, including the Indonesian archipelago and the Solomon Islands. In the latter group there is a separate subspecies on each major island.

Here is perhaps another example of birds evolving in isolation, but at the subspecies level instead of between species as in Darwin's finches. There appears to be no adaptive function in plumage changes as there is in the bill shapes of Darwin's finches (discussed elsewhere). The subspecies of golden whistlers seem to have evolved in isolation, but if they met they probably would not interbreed. This raises the question of when a subspecies becomes a full species. If races of a bird from different islands will not interbreed, then surely they are different species?

GOLDEN WHISTLER

CLASS **Aves**

ORDER **Passeriformes**

FAMILY **Pachycephalidae**

GENUS AND SPECIES ***Pachycephala pectoralis***

ALTERNATIVE NAMES
Cutthroat; thunderbird; whipbird

WEIGHT
About 1⅕ oz (35 g)

LENGTH
Head to tail: 6½–7¼ in. (16.5–18.5 cm)

DISTINCTIVE FEATURES
Male: black head, black band around upper breast, separating white throat from yellow underparts, olive-green upperparts, black or gray tail with black tip; gray and black wing tips. Female: gray-brown upperparts and gray-white underparts; buff breast.

DIET
Mostly insects; some fruit

BREEDING
Age at first breeding: 1 year; breeding season: August–December; number of eggs: 2 or 3; incubation period: about 15 days: fledging period: about 14 days; breeding interval: 1 year

LIFE SPAN
Not known

HABITAT
Forests, woodland; coastal scrub; mangroves, orchards, parks and gardens

DISTRIBUTION
South and east Australia, Java, Sonda islands, New Guinea, Solomon Islands, Fiji islands, Vanuatu archipelago

STATUS
Common

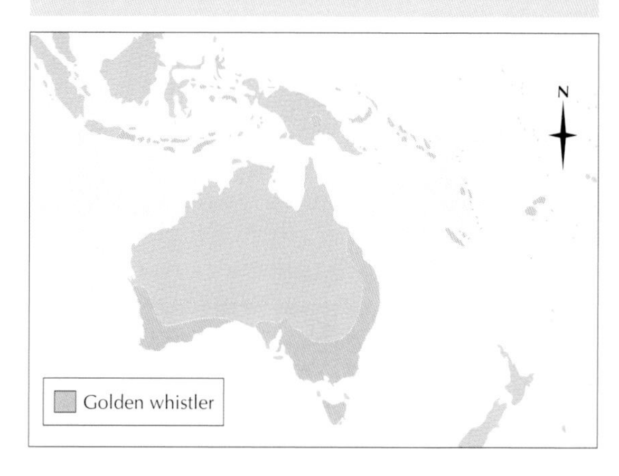

WHISTLING DUCK

A white-faced whistling duck bathes on a stretch of water in KwaZulu-Natal, South Africa. This species is about 17 inches (43 cm) long and weighs about 23 ounces (650 g).

THE WHISTLING DUCKS, ALSO known as tree ducks, are gooselike waterfowl related to the swans (discussed elsewhere), the link being the Coscoroba swan, *Coscoroba coscoroba*. They are similar to the swans in much of their behavior and differ from other ducks in their anatomy. The name whistling duck is derived from the birds' whistling cries. It is perhaps a better name than tree duck for this group because another group of ducks, the perching ducks, which includes the wood duck, *Aix sponsa*, and mandarin duck, *Aix galericulata*, spends more of its time in trees. Whistling ducks are about 17–21 inches (43–53 cm) long, with long necks, held upright, and short tails. The legs are fairly long and the feet are large. Like that of a swan, the bill has a hooked nail on the tip. Both sexes have similar plumage.

The eight species of whistling ducks all belong to the genus *Dendrocygna* and are found in tropical and subtropical parts of the Old and New Worlds. The white-faced whistling duck, *D. viduata*, occurs on both sides of the Atlantic Ocean. In America it is found from Costa Rica to northern Argentina and in Africa from the southern borders of the Sahara to the Transvaal and Madagascar. The head and neck are black and there is white on the face and throat. The back and breast are chestnut, the flanks finely striped with black and white and the underparts black. The fulvous whistling duck, *D. bicolor*, has a rather similar distribution, ranging from California to northern Argentina and throughout most of eastern Africa and Madagascar. It is also found in India. That two species of the group occur on both sides of the Atlantic is most unusual. The fulvous whistling duck is light brown with a darker back, and the flank feathers form pale stripes. The wandering whistling duck, *D. arcuata*, is darker and ranges from the Philippines and Sumatra to Australia. The plumed whistling duck, *D. eytoni*, with greatly elongated flank feathers, is another inhabitant of Australia, while the smallest of the group is the lesser whistling duck, *D. javanica*, of India, southern China and Southeast Asia.

Poorly known habits

The habits of the whistling ducks are not particularly well known because these birds are most active at night. The main stimulus to research on wildfowl has usually been their importance in

The fulvous whistling duck is the least arboreal (tree-living) whistling duck. It is a wide-ranging species. In the United States it nests in California and some southern states.

sport, but whistling ducks are not considered good sport by hunters and little research has been undertaken regarding them. Whistling ducks fly slowly, often circling around the hunters. The temptation for hunters to shoot at such easy targets is reduced further by the taste of their flesh. It is not very palatable, that of the lesser tree duck being particularly unpleasant.

For most of the year, whistling ducks live in flocks on rivers, swamps or pools. Many rest during the day in the cover of the swamps, but some, such as the black-billed whistling duck, *D. arborea*, often perch in trees. Whistling ducks' voices are shrill whistles that can be heard as they fly overhead or at night when they are feeding. In the Tropics, whistling ducks do not move far in search of food, but populations of the same species at higher latitudes migrate short distances. The American population of the black-bellied whistling duck, *D. autumnalis*, migrates to Mexico. This species, once known as the red-billed whistling duck, has a striking black belly and white flanks.

Crop eaters

Whistling ducks are mainly vegetarian, feeding on land or diving to collect food from the bottom of shallow water. Some species damage crops,

BLACK-BELLIED WHISTLING DUCK

CLASS **Aves**

ORDER **Anseriformes**

FAMILY **Anatidae**

GENUS AND SPECIES ***Dendrocygna autumnalis***

ALTERNATIVE NAMES
Black-bellied tree duck; red-billed whistling duck; red-billed tree duck; cornfield duck

WEIGHT
29 oz. (830 g)

LENGTH
Head to tail: 21 in. (53 cm)

DISTINCTIVE FEATURES
Gray head; bright red bill; ruddy breast and back; black belly; bold white stripe on upperwings; all-black underwings show in flight

DIET
Grain, grass seed, other plant material

BREEDING
Age at first breeding: 1 year; breeding season: April–October (longer season in Tropics); number of eggs: 12 to 14 (sometimes 2 broods); incubation period: 25–30 days; fledging period: 53–63 days; breeding interval: 1 year

LIFE SPAN
Not known

HABITAT
Lagoons in low-lying, lightly wooded country near cultivation; avoids deep lakes unless with extensive shallow margins

DISTRIBUTION
Northern South America east of Andes; Central America; southern North America as far north as Texas and Arizona

STATUS
Fairly common

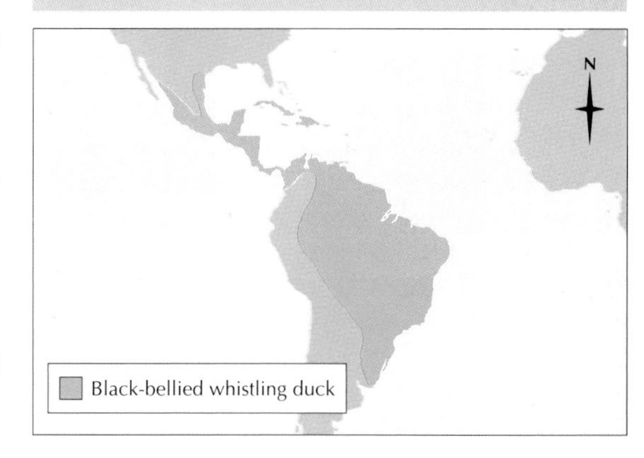

and the black-bellied whistling duck is referred to as the cornfield duck in the United States because of its attacks on corn, or maize, *Zea mays*. The ducks alight on the stalks, strip the husks and eat the grains from the cob. The black-billed whistling duck often feeds on the fruit of the royal palm, genus *Roystonea*, settling itself on the uppermost fronds.

Unlined nests

Tree ducks nest in a variety of sites. The nests usually are on the ground, hidden among reeds or grasses, the stalks of which are bent over to conceal them. However, trees and bushes may also be used as nesting places. Either a nest is built in a fork in a tree or the eggs are laid in a hollow or in the abandoned nest of a crow or other large bird. Unlike those of other ducks, whistling duck nests are not lined with down.

The partners in a pair of whistling ducks remain faithful to each other. Whereas in other ducks the male has little to do with the nesting, the male whistling duck at least shares the incubation duties, and the two preen each other, always a sign of a close bond between a pair. There are 6 to 14 eggs in a whistling duck clutch and they are incubated for 25–30 days. The male also helps in guarding the chicks.

A wide family

Although to most people ducks, geese and swans are well-defined groups of birds, ornithologists place them in a single family, Anatidae. For the most part, these three types of birds are difficult to confuse with each other. However, there are anomalies: birds or groups of birds that do not seem to fit conveniently into any of the three categories. The pygmy geese, genus *Nettapus*, and the Coscoroba swan are two examples of this. In fact, the arrangement of the Anatidae into ducks, geese and swans is not based on proper taxonomic divisions but merely on the accepted usage of the birds' common names. The whistling ducks are an example of waterfowl being called ducks when they really belong with the swans and geese in a group separate from the ducks. This becomes apparent when their appearance and behavior are closely examined: whistling ducks not only look like geese but also behave like geese, and swans for that matter. They have the same gooselike triumph ceremony (see Canada goose, discussed elsewhere), and their courtship behavior resembles that of swans and includes the dipping and neck-stretching that is typical of the latter. The male whistling duck's assistance in egg incubation is also a behavior found in some swan species.

White-faced and black-bellied whistling ducks gather on a freshwater marsh. These two species often associate where their ranges overlap.

WHITE BUTTERFLIES

A green-veined white,* Pieris napi*, feeds on a thistle head. Some white butterflies live at altitudes of up to 18,000 feet (5,500 m) or more, higher than any other butterfly.

THE 1,200 SPECIES THAT make up the family Pieridae are sometimes referred to collectively as the white butterflies, even though many pierids, especially tropical species, are brilliantly colored in yellow, orange and red. Nevertheless, there are many predominantly white butterflies among the pierids, especially in the subfamily Pierinae, which includes such species as the desert orange tip (*Anthocharis cethura*), the large white (*Pieris brassicae*) and the small white (*P. rapae*). An unusual American pierine butterfly is the pine white, *Neophasia menapia*, which occurs in forests in western North America from Canada south to Mexico and east as far as South Dakota and Nebraska. Whereas the food plants of Pieridae larvae are mostly of the mustard family Brassicaceae or the related caper family Capparaceae, the larva of the pine white, as its name suggests, feeds on the needles of conifers, including pines, genus *Pinus*, and firs, genus *Abies*. Pine-white larvae sometimes are abundant enough to do considerable damage in pine woods and plantations.

All members of the Pieridae pupate in the same way as the swallowtail butterflies (discussed elsewhere in this encyclopedia). The pupa is attached by the hooklike tail, called the cremaster, usually to a vertical surface, and has an anchoring girdle of silk midway along the body. The eggs are bottle-shaped and ribbed and stand upright on the leaves after they have been laid. The larvae are often green but show considerable diversity of form.

Crop pests

The large white and the small white are both notorious pests of cabbage and other brassica crops, on which their larvae feed. The large white generally is the less numerous of the two species, although it sometimes migrates in huge swarms. Its bristly larva is dull green and yellow with blackish mottling and is often found feeding on nasturtiums, genus *Tropaeolum*, as well as on cabbages. The larva has an unpleasant smell and is distasteful to birds, although they eagerly search for and eat the pupae during the winter. Many large white larvae are parasitized by the small ichneumon wasp *Apanteles glomeratus*.

The very abundant small white also migrates in swarms, though less frequently. Its larvae are not protected by any distasteful properties but are well camouflaged by their green color. Many also suffer from parasites. In both species the sexes are distinct, the black markings of the females being heavier and always including a pair of black spots on the forewing. Both species usually have a spring and a summer brood, the latter being more strongly marked with black in both sexes, although the timing and number of generations may vary from place to place with the length and intensity of the favorable season.

Both the large white and the small white are Eurasian butterflies, although the latter has spread widely over the world, probably by accidental conveyance of pupae. It reached Quebec in about 1860 and in the next 25 years penetrated all over temperate North America. Known in the New World as the European cabbage butterfly, it is a serious pest of cultivated plants of the mustard family. It appeared in New Zealand in 1929 and in southern Australia some years later, reaching Tasmania in about 1940. In these regions, biological control has met with some success. The agent used is *Pteromalus puparum*, a chalcid wasp (discussed elsewhere), which lays its eggs in the newly formed pupa.

There are about 30 pierine species found in North America, including three further species of the genus *Pieris*: the West Virginia white (*P. virginiensis*), which lives in parts of the northeastern and eastern United States; the mustard white (*P. oleracea*), which occurs in the northeastern United States and into Canada; and the margined white (*P. marginalis*), which is found in southwestern Canada and in the western United

States east across Wyoming and Colorado. The caterpillars of all three species feed on plants in the mustard family, the West Virginia white specializing in *Dentaria diphylla* and *D. laciniata*.

Sulphur butterflies

Besides the typical white butterflies, the Pieridae contains a subfamily of yellow, or sulphur, butterflies, the Coliadinae. In Europe two of the most familiar are the brimstone, *Gonepteryx rhamni*, the male of the which is a conspicuous sulphur yellow in color, and the clouded yellow, *Colias croceus*, the larva of which feeds on leguminous plants, including alfalfa, *Sativa medicago*. The larva of the orange sulphur, *C. eurytheme*, of North America also feeds on alfalfa, to such an extent that the adult is also known as the alfalfa caterpillar butterfly. The orange sulphur is one of a number of species that can have more than one color pattern. The usual pattern is yellowish orange with black margins. Some females, however, can be white with black borders. Another North American sulphur butterfly, the California dogface, *C. eurydice*, has officially been the state insect of California since 1972.

A third subfamily of the Pieridae is the Dismorphiinae, or mimic-whites. Most of the species in this group, many of which are brightly patterned, are confined to South and Central America. Among the exceptions are the wood white, *Leptidea sinapis*, of Eurasia, and the Costa-spotted mimic-white, *Enantia albania*, which occurs in tropical Mexico and occasionally strays into Texas. The fourth Pieridae subfamily is the Pseudopontiinae, which contains only one species, *Pseudopontia paradoxa* of Africa.

WHITE BUTTERFLIES

PHYLUM **Arthropoda**

CLASS **Insecta**

ORDER **Lepidoptera**

FAMILY **Pieridae**

GENUS **Many, including *Anthocharis, Neophasia, Pieris***

SPECIES **About 1,200, including *A. cethura, N. menapia, P. rapae***

LENGTH
Wingspan: ¾–2¾ in. (2–7 cm)

DISTINCTIVE FEATURES
Medium-sized butterflies; mostly white, orange or yellow in color; 6 walking legs

DIET
Host plants vary according to species

BREEDING
Eggs generally laid singly on appropriate host plants

LIFE SPAN
Not known

HABITAT
Varies according to species, including tropical and temperate woodland, grasslands, marshes, scrub, cultivated crops

DISTRIBUTION
Worldwide, range varying according to species

STATUS
Common

Butterfly pigments

In many of the most brilliant butterflies and moths the colors are produced by the effect known as structural coloration (see "Urania moth" article). In most of the Pieridae, the bright colors—red, orange and yellow—as well as white are due to chemical pigments produced by the insect and deposited in the scales of its wings during development. These pigments are known as pterines. The white substance in the scales of the common white butterflies is a pterine called leucopterine, which chemically resembles uric acid and was formerly confused with it. Different kinds of butterflies are colored by different types of pigments. The scales of the wood white, for example, contain a flavone, or anthoxanthin (a pigment of a different nature from a pterine). This pigment is also found in the marbled white, *Melanargia galathea*, a member of the Satyridae family of butterflies.

Caterpillars of the large white,* Pieris brassicae, *munch their way through cabbage leaves. A Eurasian species, the large white is a rare vagrant or escapee in North America.

WHITE-EYE

A Javan white-eye,* Zosterops flavus, *takes to the wing. White-eyes show a wide variation in plumage, particularly among the species living on islands, and are popular cage birds.

As is the case with several bird names, the term white-eye can refer to more than one type of bird. Several of the pochard group of ducks, tribe Aythyini, are called white-eyes. However, the white-eyes considered in this article are small birds of the family Ploceidae that are found in most of the Old World Tropics from West Africa to the Pacific islands, and from Japan to the sub-antarctic Macquarie Island in the Southern Ocean.

Most of the 94 species of white-eyes are classed in the genus *Zosterops*, which sometimes doubles as a common name. White-eyes generally are about 4–5 inches (10–13 cm) long, and the bill usually is slender and very slightly curved, the wings rounded and the tail square. Both sexes of white-eyes are similar and tend to be green or yellowish above and yellow or gray below. Those living on islands often lack yellow in the plumage.

As the name suggests, a principal characteristic of this group of birds is the white ring of minute feathers around the eye, but this feature is highly variable. In some African species the ring is a large patch, while in some other white-eyes it is missing entirely. The yellow-spectacled white-eye, *Z. wallacei*, also known as Wallace's white-eye, native to the Lesser Sunda Islands, has a yellow eye ring, and the olive black-eye, *Chlorocharis emiliae*, a species of white-eye from Borneo, has a black eye ring.

Among the best known of the white-eyes is the African yellow white-eye, *Z. senegalensis*, which ranges from Senegal to South Africa. It exists in 17 forms and is also known as the green or the yellow white-eye, depending on the brilliance of its plumage. The most common Australian white-eye is the silver-eye, *Z. lateralis*, which is often a pest in orchards and suburban gardens. It is olive green above and gray underneath and lives in southern and eastern Australia and on Tasmania.

Variation with climate

White-eyes live in flocks of up to 100 in a very wide variety of wooded habitats, ranging from dense forests, where they are found only on the margins, to scrub. They occur in the Australian coastal mangrove swamps and in the wooded highlands up to the tree line. Often a species is restricted to a single habitat type, with different white-eyes living close together, but not overlapping, as the altitude increases up a mountainside or as forest changes from dry to wet. The ornithologist R. E. Moreau demonstrated that the white-eyes of Africa varied from one environment to another in accordance with several biological rules. Bergmann's rule, for instance, states that the body size of an animal becomes larger toward the cooler part of its range, an adaptation to the conservation of heat. This rule applies in the African white-eyes. Their wing lengths also increase with altitude, and their plumage becomes darker in areas of higher humidity.

Calling while feeding

Flocks of white-eyes keep in touch by means of quiet calls while they forage among bushes and trees in a straggling procession. As one bird flies from a tree, calling, others follow in a stream. White-eyes eat insects, fruit and nectar. The insects are found among the trees and include aphids, caterpillars and flying termites. Like the honeyeaters (discussed elsewhere), white-eyes have brush-tipped tongues, which they use for mopping up nectar or fruit pulp. To feed from

AFRICAN YELLOW WHITE-EYE

CLASS **Aves**

ORDER **Passeriformes**

FAMILY **Zosteropidae**

GENUS AND SPECIES ***Zosterops senegalensis***

ALTERNATIVE NAMES
Green white-eye; yellow white-eye

WEIGHT
⅓ oz. (10 g) on average

LENGTH
Head to tail: 4⅓–4¾ in. (11–12 cm)

DISTINCTIVE FEATURES
Small; thin bill; greenish yellow above, bright yellow below; white eye ring

DIET
Insects; nectar, fruit

BREEDING
Age at first breeding: 1 year; breeding season: August–February; number of eggs: 2 to 4 (3); incubation period: 11 days; fledging period: 14 days; breeding interval: 1 year

LIFE SPAN
Not known

HABITAT
Evergreen and riverine forests; eucalyptus plantations; gardens

DISTRIBUTION
Much of sub-Saharan Africa

STATUS
Common

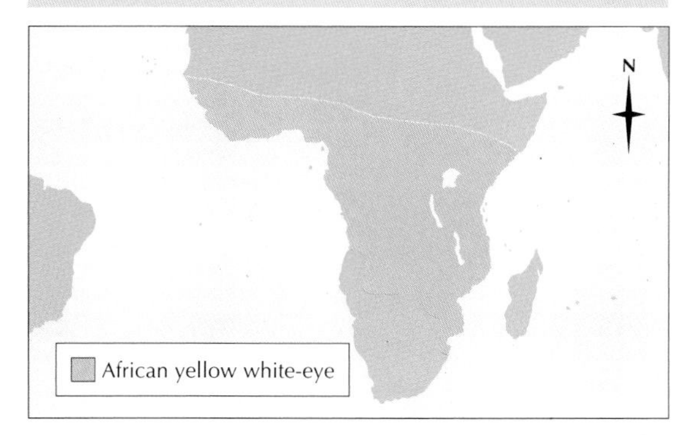

flowers, they pierce the base with their pointed bill and then lick out the nectar with the tongue. In the case of soft fruits such as pawpaw, figs and grapes the birds pierce the skin and extract the juice and pulp. Because the white-eyes remove only a small part of each fruit, the damage they do to a crop is out of all proportion to the amount the birds eat. The flocking habits of the white-eyes further increase the amount of destruction they cause.

Family similarity

For all their variety in plumage, white-eyes have very similar nesting habits. At the beginning of the breeding season, the flocks split up and the males sing a melodic, far-carrying trill. The nest is a deep cup, about 2 inches (5 cm) across, slung between two twigs. It is made of grasses or lichen, bound with cobweb and lined with finer grasses, kapok or sheep's wool. Both parents incubate the two or three eggs for 10–12 days. The chicks stay in the nest for 9–13 days and are fed on caterpillars, pulped by passage through the parents' bills. After they have fledged, the chicks stay with their parents for 2–3 weeks.

Successful colonists

Most species of white-eyes do not migrate, although part of the Tasmanian population of the silver-eye travels across the Bass Strait to New South Wales. Even so, the white-eyes have been extremely successful at colonizing islands. They were introduced to one of the Hawaiian islands and have since spread around the archipelago. In about 1856, Tasmanian silver-eyes arrived in large numbers on the western coast of North Island, New Zealand, a distance of some 1,200 miles (1,900 km) from their starting point. Then they rapidly colonized the whole of the country. Perhaps more remarkable still is their colonization of the subantarctic Macquarie Island, about 700 miles (1,100 km) south of New Zealand.

An oriental white-eye,* Z. palpebrosa*, rests after a meal, its head covered in pollen. This species ranges from Afghanistan to Indonesia.

WHITEFLY

WHITEFLIES ARE NOT FLIES at all but extremely small bugs very closely related to the aphids and scale insects (both discussed elsewhere). They are rarely more than ⅛ inch (3 mm) long and many are smaller. The adults resemble minute white moths as they fly around. The color is produced by a fine waxy powder that covers the body and wings. This substance is given out from the anus, which lies in a cavity overlaid by a tongue-shaped flap, or lingula. The cavity itself is covered by an operculum (lid). Originally, the wax probably was given out as a waste substance, as is the case in some other bugs, but in whiteflies this waste is put to use, possibly as a protective covering.

The majority of known whiteflies are tropical, and they would largely escape notice but for the fact that some are serious pests of crops. It is highly likely that the majority of living whitefly species have yet to be discovered. Among the better-known species are the cabbage whitefly *Aleyrodes proletella,* which infects cabbages in Britain and other temperate countries, and the greenhouse whitefly, *Trialeurodes vaporariorum,* which is a tropical insect that establishes itself in greenhouses, attacking tomatoes, cucumbers and many other plants. It is believed to have come originally from Brazil. A third well-known species, the citrus whitefly, *Dialeurodes citri,* is a pest of citrus fruits in the southern United States.

Harmful sapsuckers

Young cabbage whiteflies congregate on the undersides of the leaves of cabbages, genus *Brassica,* where they breed throughout the summer. If they are disturbed in the fall, the adults rise in clouds before settling again like tiny snowflakes. This species, as well as the citrus and greenhouse whiteflies, discolor and weaken the plants they infest by sucking their sap, in the same way as do aphids. Also like the aphids, they cover the plants with sticky honeydew. On citrus fruit, the honeydew promotes the growth of a fungus, the sooty mold *Meliola camelliae,* which so discolors the fruit that even though it is not spoiled altogether it has to be expensively washed for marketing. The reason whiteflies and other bugs take in so much sap, in excess of their need for carbohydrates, is that they also need nutrients such as amino acids, which are not highly concentrated in the liquid.

Eggs laid in circles

Whiteflies lay their eggs on the undersides of leaves, sometimes in a double layer. They are usually arranged in an arc or a circle, because as she lays, the female keeps her head in one place and moves her body around the axis thus provided. Each egg has a stalk and is laid upright, a batch of eggs looking rather like a group of minute pegs. The stalks of the eggs

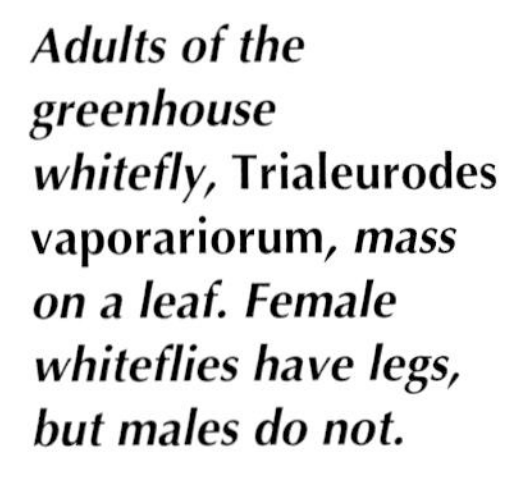

Adults of the greenhouse whitefly,* Trialeurodes vaporariorum, *mass on a leaf. Female whiteflies have legs, but males do not.

WHITEFLIES

PHYLUM **Arthropoda**

CLASS **Insecta**

ORDER **Hemiptera**

SUBORDER **Homoptera**

FAMILY **Aleyrodidae**

GENUS **Many, including *Aleyrodes, Dialeurodes, Trialeurodes***

SPECIES **Many, including cabbage whitefly, *A. proletella*; citrus whitefly, *D. citri*; greenhouse whitefly, *T. vaporariorum***

LENGTH
Wingspan: up to ⅕ in. (5 mm)

DISTINCTIVE FEATURES
Tiny insect, resembling minute moth; covered with fine, white, waxy powder

DIET
Plant sap

BREEDING
Eggs laid on underside of leaves; 3 nymphal stages and pupal stage precede adult stage

LIFE SPAN
Usually a few weeks

HABITAT
Underside of leaves of host plant

DISTRIBUTION
Most abundant in Tropics; also temperate regions

STATUS
Common

pierce the outer layer of leaf cells and, because they are hollow, they draw moisture from the leaf by capillary action. As long as the plant is well supplied with water, development of the eggs proceeds normally. If the leaf withers, however, the development of the eggs is arrested, and if the plant wilts during a drought, the egg goes into a resting stage until more sap is available. Sometimes, whitefly eggs go into a resting stage for no obvious reason. All these are almost certainly adaptations to life in the Tropics, where droughts are more likely to occur.

Whitefly larvae are tiny and scalelike. Initially, they move about the host plant, feeding on the sap and excreting honeydew. After a time, they cast their skins, their legs degenerate and they become static, developing protective rods or filaments of wax around themselves. However, they continue to feed. A further molt and a third nymphal stage follow before the larvae settle down into a state resembling pupation. The larvae carry on feeding for a while, then stop while the adult appendages develop. After a period of rest, which varies according to species, the winged insects emerge fully developed.

The characteristic way in which bugs develop is by a series of instars, or stages separated by skin changes. In the course of these, the insect gradually takes on the characteristics of the adult, while continuing to feed and move about. This development contrasts with, for example, the pupation of a moth. The so-called false pupation of the whiteflies is a departure from the normal way in which bugs develop, and it closely parallels the true pupation of such insects as butterflies, bees and beetles. The differences are that in the false pupation, the insect has much the same shape as the previous larval stage and it keeps feeding, at least to begin with.

Imported killers

The greenhouse whitefly can be effectively controlled by the use of a minute chalcid wasp species, *Encarsia formosa*, which is a parasite of the whitefly. The males of *E. formosa* are rare and apparently do not mate even when they occur. The parasite reproduces by parthenogenesis; that is, by the development of unfertilized eggs. The female wasps lay their eggs in the larvae, or scales, of the whiteflies, and the scales that are parasitized turn black. To introduce the wasp, bunches of tomato leaves bearing parasitized scales are obtained and suspended in the greenhouse. Blackening of the whitefly scales is generally noticed 2–3 weeks after the introduction of the parasitized material. A nocturnal temperature of about 55° F (13° C) is needed to maintain the wasps.

An adult whitefly with larval and pupal scales. The black scales have been parasitized by a chalcid wasp, which has laid its eggs in them. Its larvae hatch in the eggs and feed on their hosts.

WHITING

One characteristic of the whiting is that the long anterior anal fin begins directly opposite the center of the frontmost dorsal fin and ends level with the posterior base of the second dorsal fin.

A MEMBER OF THE COD family (Gadidae), the whiting is an important food fish. It reaches a length of 28 inches (70 cm), the females being slightly larger than the males. The whiting can be distinguished by its silvery sides and belly and by the dark blotch at the base of each pectoral fin. Young fish also have a very small, inconspicuous chin barbel, which has vanished by the time the adult stage is reached. The whiting's snout is long and rather pointed, and the upper jaw is longer than the lower. There are three soft-rayed dorsal fins and two anal fins. The fish's coloration varies on the back and the head, from yellowish brown to green to dark blue, but all whiting have silvery sides and a silvery belly. The whiting occurs in the northeast Atlantic, ranging from the North Sea off Iceland and Norway through the Irish Sea and the English Channel and down the coasts of France, Spain and Portugal. It is also found in the Mediterranean and Black Seas.

Abundant throughout its range

The whiting is very common throughout its range, being particularly abundant in the North and Irish Seas. Unlike most members of the cod family, it lives mainly in shallow waters. It is most common from 100–330 feet (30–100 m) and rarely goes below 4,000 feet (1,220 m). Sometimes whiting come close inshore and are caught in just a few feet of water. Like all members of the cod family, the whiting makes seasonal migrations within its range. A strong fighter for its size, the whiting makes a good sporting fish for anglers. The record for a whiting caught on rod and line off the British coast is almost 7 pounds (3.2 kg).

Whiting feed mainly by day, mostly in midwater or just off the bottom. Their prey consists principally of smaller fish, such as sprats, younger whiting, herring and sand eels, but they also eat large quantities of crustaceans, such as shrimps, prawns and crabs, and occasionally take cuttlefish, small squids and worms. Young whiting, which generally live close to the shore, feed on shrimps, young shore crabs, amphipods, small gobies and sand eels.

Shallow-water spawner

In northern areas, whiting spawn mainly in April and May, but they may continue into July. In the south they may begin as early as January. Spawning takes place principally in depths of

WHITING

CLASS **Osteichthyes**

ORDER **Gadiformes**

FAMILY **Gadidae**

GENUS AND SPECIES ***Merlangius merlangus***

WEIGHT
Up to about 7 lb. (3.2 kg)

LENGTH
28 in. (70 cm)

DISTINCTIVE FEATURES
Elongated with small, pointed head; small chin barbel in young, absent in adult; 3 dorsal fins, 2 anal fins. Coloration variable: upperparts yellowish brown, dark blue or green; sides and belly white and silvery; small, dark blotch at upper base of each pectoral fin.

DIET
Invertebrates and small fish

BREEDING
Breeding season: January–July, but mostly in spring; number of eggs: up to 285,000

LIFE SPAN
Up to 10 years

HABITAT
Shallow inshore waters, mainly over mud and gravel but also over sand and rock

DISTRIBUTION
Northeast Atlantic south to Portugal; also Black and Mediterranean Seas

STATUS
Not threatened

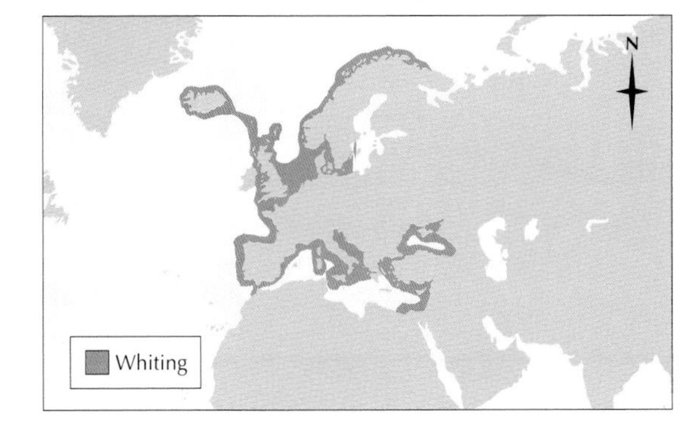

300 feet (90 m) or less, and sometimes in only 60 feet (18 m) of water. The whiting's relatives, the cod (*Gadus morhua*) and the haddock (*Melanogrammus aeglefinus*), spawn in deeper waters. The whiting's eggs resemble those of the cod and haddock but are smaller, averaging about 1 millimeter in diameter. They are laid in great numbers and float at the surface. When it hatches, the young larva is about 2 millimeters long and drifts at the surface feeding on plankton. This is the most critical stage of the fish's life, as it is at the mercy of numerous predators. If it survives, the young fish grows fairly quickly, reaching about 6 inches (15 cm) on average in its first year, increasing to about 9 inches (23 cm) in its second year, when it matures, and to 12 inches (30 cm) in its third year. Whiting live up to 10 years.

Whiting spend their early months of life in shallow water, migrating to the open sea when they are about 1 year old.

Deadly shelter

Very young whiting, once they have reached a length of about ¾ inch (2 cm), sometimes associate with jellyfish, particularly those of the genus *Cyanea*. The fish shelter under the jellyfish's bell in small shoals and swim among the stinging tentacles, enjoying the protection without apparently coming to any harm themselves. Once they reach 2½ inches (6.4 cm) in length, whiting migrate to the bottom and are found in great numbers close to the shore, especially in sandy bays and estuaries.

As mentioned above, the most perilous stage of the whiting's life is when it first hatches and is drifting at the surface. At this time it is vulnerable to seabirds, predatory fish and adverse tides and currents. Although a female whiting produces up to 285,000 eggs at a single spawning, very few reach adulthood. Around Britain, particularly during July and August, young whiting frequent the shrimping grounds off the northwest English coast, where thousands are killed each year through the operations of shrimpers.

WHYDAH

The whydahs form a group of sparrowlike birds that, together with the indigobirds, make up the family Viduidae. Traditionally, the whydahs were considered to be part of the family Estrildidae, comprising the waxbills and their relatives, but recent genetic studies show them to be more closely related to each other and to warrant a family of their own. Whatever the true relations of the whydahs, they form a distinct group of African, seed-eating birds that lay their eggs in the nests of weaver finches.

There are 19 species in the whydah and indigobird family, four called whydah and five bearing the name paradise-whydah. The females are drab and sparrowlike throughout the year but in the breeding season the males have a special plumage, which usually is shiny black, sometimes with very long tail feathers. The village indigobird, *Vidua chalybeata*, is an example of a whydah without a long tail. In the breeding season, the male of this species resembles a small black finch, but the feathers have a bluish gloss and the bill is white. The male of the pin-tailed whydah, *V. macroura*, which is found throughout most of Africa south of the Sahara, measures 12–13 inches (30–33 cm), of which up to 8 inches (20 cm) is made up by the four central tail feathers. The breeding plumage is black and white, the tail, back and crown being glossy black, and the bill is red. Another distinctive and widespread whydah is the paradise-whydah, *V. paradisaea*, which has a 10-inch (4-cm) tail though the two outer feathers are only half the length of the inner two feathers. A breeding male has striking plumage: a black head contrasting with a yellow nape, chestnut breast band, bright yellow lower breast and belly and black upperparts.

Birds of open spaces

Whydahs live in the savannas and plains of Africa. Like the weavers, however, they form large flocks outside the breeding season, sometimes mixing with flocks of weavers and weaver finches. Some species are also found around towns and villages. Whydahs are predominantly seed-eaters, but they also take insects. They feed mainly on the ground, searching for small seeds by scratching with their feet.

Cuckoolike habits

For a long time, ornithologists remained uncertain about the breeding habits of the whydahs and only a very few nests were ever described. It was suspected that these descriptions were based

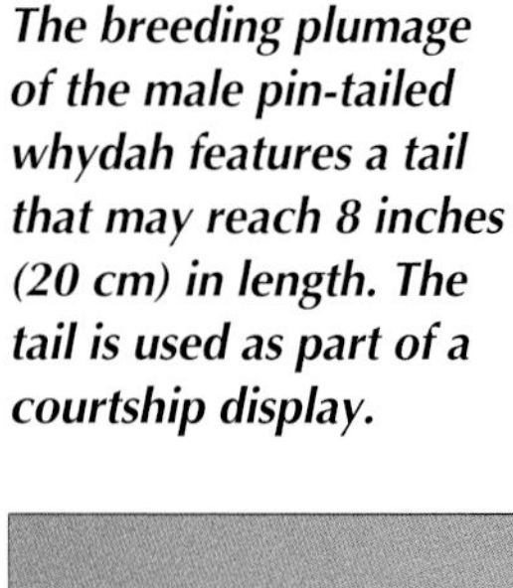

The breeding plumage of the male pin-tailed whydah features a tail that may reach 8 inches (20 cm) in length. The tail is used as part of a courtship display.

PIN-TAILED WHYDAH

CLASS **Aves**

ORDER **Passeriformes**

FAMILY **Viduidae**

GENUS AND SPECIES ***Vidua macroura***

WEIGHT
7/10–8/10 oz. (19–22 g)

LENGTH
Head to tail: 12–13 in. (30–33 cm) (breeding male); 4½ in. (11.5 cm) (female)

DISTINCTIVE FEATURES
Breeding male: four black central tail feathers up to 8 in. (20cm) long; white underparts; white wing patches; white cheeks; black crown; red bill. Nonbreeding male and female: off-white underparts; brown upperparts; black-and-white stripes on head. Nonbreeding male: red bill. Female: brown bill.

DIET
Mostly seeds; insects

BREEDING
Age at first breeding: 1 year; breeding season: mostly September–January (Africa), April–November (Puerto Rico), but egg-laying varies according to region; number of eggs: up to 26; incubation period: 10–12 days; fledging period: about 14 days; breeding interval: 1 year

LIFE SPAN
Not known

HABITAT
Savanna, grassland, scrub; parks, gardens

DISTRIBUTION
Sub-Saharan Africa; successfully introduced to Puerto Rico

STATUS
Common

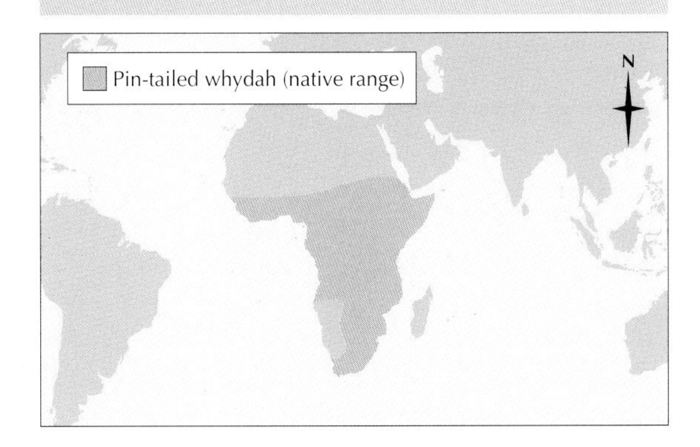

Pin-tailed whydahs are bold and active birds that favor dry, open bushed and wooded habitats.

on mistaken identifications. This proved to be the case and it was later shown that whydahs do not build nests. They are parasites, like cuckoos, and lay their eggs in the nests of their near relatives, the weaver finches. The whydahs' parasitic habits were discovered in 1907.

In the whydahs, there is no bond between male and female, which is hardly surprising considering that when mating is completed the male's role is ended. The males are polygamous and are aggressive toward each other. Many have elaborate courtship displays, rivaling the birds of paradise (discussed elsewhere) in their behavior as well as in their plumage. The male paradise-whydah clears a patch of ground about 4 feet (1.2 m) across and prances about on it to attract females. Others, such as the pin-tailed whydah, perform display flights, ascending 100–200 feet (3–60 m) and then fluttering slowly down with the long tail waving.

Also known as the queen whydah, the shaft-tailed whydah,* V. regia, *is distinguishable by the slight flare to the end of its tail and by its straw-colored underparts.

Skillful mimic

The whydahs possess a remarkable mechanism that enables them to ensure their offspring are successfully brought up by the foster parents. This is necessary because of the manner in which the young weaver finches are reared. They are fed by the parents' thrusting their bills into the chicks' mouths and regurgitating semidigested food. To guide the parents, each chick has a pattern of black markings in the mouth and bright spots around the bill. They also have special begging calls. The markings and calls vary between the species, and a weaver finch feeds only chicks with the correct markings and calls.

It is essential that the whydah chicks should be able to mimic the recognition signals of their foster parents' chicks in order to be accepted and fed. Each whydah species has evolved the mouth markings and begging calls of its particular host, and the fledgling birds have plumage resembling that of the host fledgling.

Because the whydah chick needs to resemble the host chick, it is essential for the female whydah to lay her eggs in the nests of the correct weaver finches. This happens with cuckoos whose egg markings mimic those of the host species, and it seems that the female cuckoo somehow is able to recognize nests of the species in which she herself was raised. In the whydahs, it seems that the male also plays a part in ensuring that the eggs are laid in the right nest.

There are several very similar races of paradise-whydahs, some of which live in the same areas. The parasite chick does not evict the host young, but it outcompetes them for food and usually no host young survive to fledge. Each race parasitizes a different weaver finch, and it is therefore essential for female paradise-whydahs to mate with the correct race of male, otherwise the hybrid offspring will not mimic the host. Recognition is based on song. In an outstanding example of imitation, male paradise-whydahs mimic the songs of their hosts, so that, for example, a male narrow-tailed paradise-whydah sings a song that is almost indistinguishable from the song of the melba finch, *Pytilia melba*, that reared it. The song attracts female paradise-whydahs that were also reared by melba finches and are conditioned to lay their eggs in melba-finch nests.

Each species of whydah usually lays its eggs in the nests of a particular species of weaver finch. Usually only one egg is laid in each nest, and the host's eggs are not destroyed, but if a second whydah lays in a nest where there is already one whydah egg, she will destroy one of the weaver-finch eggs. Both the host's and the parasite's eggs are white, but the whydah can recognize and destroy the host's eggs.

The 10 species of indigobirds parasitize firefinches and twinspots. Paradise-whydahs are larger and their hosts are pylylias. The whydahs parasitize waxbills.

WILD ASS

Asiatic wild asses are more lightly built than their African counterparts. This animal was photographed in the Hay-Bar Biblical Reserve in Israel.

THREE SPECIES OF WILD horses are regarded as asses. In northeastern Africa there is the African wild ass, *Equus asinus*. Asia, meanwhile, has the kiang, *E. kiang*, and *E. hemionus*, which is divided into two subspecies, the kulan (*E. h. kulan*) and the onager (*E. h. onager*). Asses have been domesticated from early times, and the domestic ass, or donkey, probably came from the now-extinct North African race, *E. a. asinus*.

Asses stand 3¼–5 feet (1–1.5 m) high at the shoulder and have a gray, rufous or brownish coat, sometimes with dark markings on the back and shoulders in the shape of a cross. The hair is lighter on the muzzle, flanks and belly. The coarse, erect mane lacks a forelock, and the tail has a tuft of long hair at the tip. The legs are sometimes striped like those of a zebra, and the ears are long.

The African wild ass is now restricted to Ethiopia and Somalia. Its numbers have been on the decline for centuries. The North African race became extinct in Roman times, and the others have interbred with feral donkeys. The Asiatic wild asses are more numerous. The kulan and onager are found across much of Central Asia, while the kiang occurs only in Tibet. Wild asses are protected in several parts of Asia. Because of this, the Asiatic species are better known than the African wild ass, so it is the life history of the former that is described here.

Animals of plain and steppe

Asiatic wild asses live on broad open plains and high steppes, penetrating high into mountains. They prefer hilly areas, but come down into the valleys in order to avoid duststorms or snowstorms. Asses avoid soft sand because it makes travel difficult and food is scarce in such areas. At one time the range of the Asiatic species was considerable. In the 17th century, it extended from the Black Sea in the west to the Yellow River of northern China in the east.

Very wary animals, wild asses cannot be approached easily, and they keep well away from human habitation even when suffering from thirst. Despite this, Russian biologists have been able to gather a considerable amount of information on their habits. The asses live in troops of 10 to 20 animals, a troop consisting of a dominant stallion, several females and some juveniles. In the fall and winter, troops may band together into herds 200 or 300 strong. Such a concentration is possible in the winter because it is then that the steppe plants flourish. In summer the population spreads out, and water holes become very important in determining the distribution of the troops. At this time they never move more than 7–8 miles (11–13 km) from water, whereas in winter, when the vegetation becomes lush, they may move six times as far away, as they can obtain water from their food.

The onager, shown here, was domesticated in ancient times, but the sturdier donkey was eventually preferred as a working animal.

The Russian researchers also discovered that the asses are active throughout the day in spring, but as the weather gets warmer they take to lying up in thickets during the hottest part of the day, emerging to graze and drink at night when the temperature has dropped.

A food for each season

The wild ass diet varies throughout the year. In spring the main foods are grasses and sedges, but as the vegetation begins to wither and dry up, the asses turn to herbs such as tansy, genus *Tanacetum*. In the northern and mountainous parts of their range, asses may be seriously affected by snowfall. If the snow is much more than 1 foot (30 cm) deep, they are severely hampered both in movement and in feeding, and they have to browse on tamarisk, genus *Tamarix*, and other bushes. The severe winters of 1879 and 1891 are thought to be the reason for the extinction of wild asses in Kazakhstan.

The rut (mating season) takes place during the spring or summer, the exact period varying from region to region. The stallions become very excited, racing around their troops, rolling on their backs and fighting each other before mating with the females. The foals are born 11 or 12 months later. The mares generally breed only once in 2 years, starting when they are 2 or 3 years old, although they can produce a foal each year if they are in good health. A few months after foaling, some of the mares and their offspring are driven from the troop to be taken in charge by younger, solitary stallions.

Stallion defends against predators

Humans have been the wild ass's main predator, due to the fact that its flesh and hide are much sought after by the inhabitants of the steppes

WILD ASSES

CLASS **Mammalia**

ORDER **Perissodactyla**

FAMILY **Equidae**

GENUS AND SPECIES **African wild ass, *Equus asinus*; kiang, *E. kiang*; kulan, *E. hemionus kulan*; onager, *E. h. onager***

ALTERNATIVE NAME
Burro (*E. asinus* only)

WEIGHT
440–880 lb. (200–400 kg)

LENGTH
Head and body: 6½–8¼ ft. (2–2.5 m); shoulder height: 3¼–5 ft. (1–1.5 m); tail: 12–20 in. (30–50 cm)

DISTINCTIVE FEATURES
Gray, rufous or brownish coat, sometimes with crosslike dark marking on back and shoulders; mane sparse and erect; tail tufted

DIET
Plant matter

BREEDING
Age at first breeding: 1 year (female), 2 years (male); breeding season: year-round but may concentrate in wet season; number of young: 1; gestation period: about 1 year; breeding interval: 1 or 2 years

LIFE SPAN
Up to about 45 years

HABITAT
African wild ass: stony deserts; kiang, kulan and onager: broad open plains, high steppes

DISTRIBUTION
African wild ass: Ethiopia, Somalia; kulan, onager: Central Asia; kiang: Tibet

STATUS
African wild ass: critically endangered; kulan: near threatened; onager: endangered

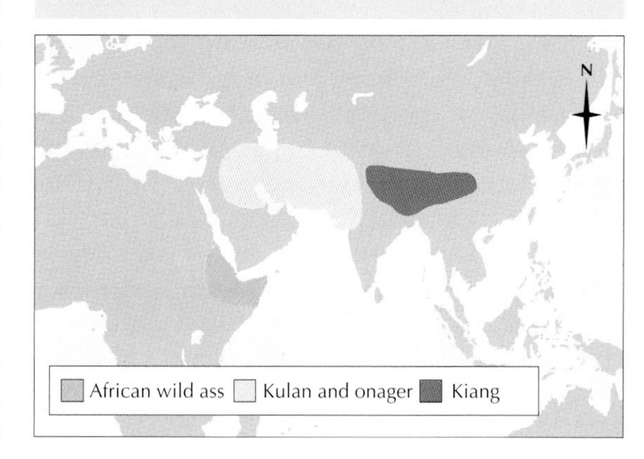

and plains. The introduction of firearms must have been a major factor in the animal's decline, as natural wariness no longer ensured its safety. Beside humans, wolves are the ass's principal predator. If danger threatens, the dominant stallion leads the troop away from the source but will turn back to chivvy the females if they lag behind, nibbling them, neighing and shaking its head. A troop of asses has little to fear from wolves, but a solitary animal becomes easy prey.

Crossbreeding between species

In simple terms, a species is defined as a group of animals that cannot breed with another group, but in all definitions there are exceptions. In the case of the horse family the most familiar exceptions to this rule are represented by the mule and the much rarer hinney. A mule is the offspring of a female horse, *E. caballus*, and a male ass, whereas a hinney's parents are a female ass and a male horse. Both crosses are nearly always sterile and cannot themselves breed, each individual having to be produced from a cross between the two parent species. The mule is very asslike, with its long ears, thin legs and small hooves, whereas the hinney is a small, horselike animal, less sturdy than the mule.

The mule has been used as a beast of burden since ancient Greek and Roman times, being an extremely robust animal and, unlike a horse, able to recover quickly from being worked to the limits of its endurance. The mule's tougher hide makes the animal less susceptible to saddle sores and chafing, and the mule is relatively insensitive to disease. All these qualities make the mule an ideal pack and draft animal, especially in desert and mountainous regions, where its ability to live on coarse herbage and very little water rivals, if not exceeds, that of the camel. Although the mule has now been largely replaced by motor transport, it is still used by armies operating in mountains.

The wisdom of the mule

Other attributes of the mule are its intelligence and its proverbial stubbornness. Mules were extensively used by the armies of World War I for hauling transport wagons and light artillery. British soldiers who worked with them spoke of them sardonically as their long-faced friends. It was said that nothing could equal the obstinacy of a mule. If a mule refused to carry out a task, there was little to be done except wait for the animal to change its mind because it would not be driven. However, a mule's obstinacy is not always blind. On one occasion a transport column was moving along a road in the summer of 1918, when the British army was advancing rapidly on the Western Front. Suddenly the leading mules stopped dead and all the rest followed suit. Their drivers were unable to make them take another step. Then, a hundred yards or so ahead of the column, a battery of field guns opened fire. The mules went forward of their own volition as soon as the firing ceased. It transpired that the battery had pulled into position during the night and was so well camouflaged that the drivers of the mules had no idea of its presence. Seemingly, though, their animals were aware that it was there.

A troop of kiangs gallops through the Paryang Valley, western Tibet. Asiatic wild asses are fast runners and can reach speeds of up to 40 mph (64 km/h).

WILD BOAR

A wild boar makes its way across a snowy landscape in France. The wild boar's bristly coat gets thicker in winter to provide protection from cold.

An ancestor of the domesticated pig, or hog, the wild boar is gray to dark brown in color, with a coat of stiff bristles. Some individuals have side whiskers on the cheeks, a slight mane, or both. The wild boar has a head-and-body length of 36–72 inches (90–180 cm), a tail up to 1 foot (30 cm) long, and a height at the shoulder of 20–43 inches (50–110 cm). It weighs 88–770 pounds (40–350 kg), and its tusks may be 12 inches (30 cm) in length, including their continually growing root.

The family party

For most of the year, the social unit of the wild boar is the family party, but in the fall family groups come together to form bands of up to 50 females and youngsters, the old males mainly remaining solitary. Wild boars live in a wide variety of habitats but prefer terrain with cover such as open woodlands, especially where there are mud wallows in which they will spend many hours at a time if undisturbed. They also make crude shelters by cutting long grass and then crawling under it, lifting the grass up so that it becomes entangled with the tall herbage around it to form a canopy. Quick-footed and good swimmers, wild boars normally avoid combat but react vigorously when provoked, slashing with their tusks.

Nocturnal rootings

Wild boars spend the hours of darkness rooting for anything edible and may travel up to 50 miles (80 km) in a night. Their diet is varied and includes acorns and beech mast, bulbs, tubers and roots of various kinds, even the roots of ferns that few other animals eat. They have a natural tendency to dig for the potato-shaped fungi known as truffles, genus *Tuber*, which are a great and expensive delicacy, especially in French and Italian cuisine. Wild boars also eat invertebrates, small vertebrates, such as mice and voles, and carrion. They may become a nuisance among cereal crops and root crops, such as beet or turnip, and among potatoes. Wild boars have been hunted for centuries, partly because of the destruction that they cause but also for their flesh and for sport.

WILD BOAR

CLASS **Mammalia**

ORDER **Artiodactyla**

FAMILY **Suidae**

GENUS AND SPECIES ***Sus scrofa***

ALTERNATIVE NAME
Wild pig

WEIGHT
88–770 lb. (40–350 kg)

LENGTH
Head and body: 36–72 in. (90–180 cm); shoulder height: 20–43 in. (50–110 cm); tail: 12 in. (30 cm)

DISTINCTIVE FEATURES
Dark brown or gray coloration; body covered in stiff bristles; 4 tusks; sometimes side whiskers and small mane; young are striped

DIET
Bulbs, tubers, vegetation, nuts, seeds, fungi; invertebrates; small vertebrates; carrion

BREEDING
Age at first breeding: about 18 months (female), about 5 years (male); breeding season: year-round, but may peak around rainy season; number of young: usually 4 to 8 (1 to 12); gestation period: 100–142 days; breeding interval: 1 year

LIFE SPAN
About 10 years

HABITAT
Varies widely, but prefers places with cover, such as forests and reed beds

DISTRIBUTION
Much of continental Europe and Asia, North Africa; also New Guinea

STATUS
Abundant in much of range

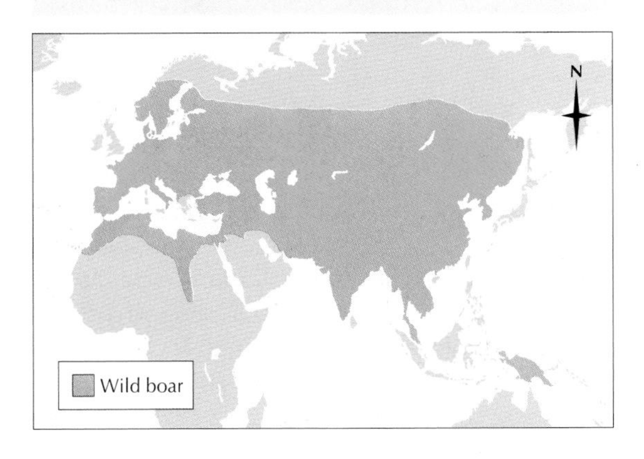

A litter of wild boar piglets suckling. The sow gives birth in a nest of vegetation that she has built. She remains there with her young for 1–2 weeks, leaving them only rarely.

Striped young

The sow produces a litter of 1 to 12 (usually 4 to 8) after a gestation period of 100–142 days. She has 8 to 14 teats, and because each piglet takes a teat at feeding time, the weaklings in a large litter may die. Piglets suckle for about 12 weeks before being completely weaned onto solid food, which they find by rooting around for themselves, although they never stray very far from the protection of their mother. The male takes no part in caring for the young, which are striped, like those of other wild pigs. Wild boars become sexually mature at 18 months, and although females may breed from about that age, males usually have to wait until they are large enough to compete for mates, which is not until they are close to full size at about 5 years of age. The animals live to about 10 years in the wild but have survived for nearly 30 years in captivity.

Early domestication

Wild boars cannot be readily herded, but they take well to life in sties or in houses, so scientists can be fairly sure that their domestication came about when humans ceased to be purely hunters and gatherers of wild food and some settled down to agriculture. Another clue is in the taboo on eating pig flesh, which seems to have originated among nomads, who had a contempt for the settled agrarian communities and expressed their views through an apparent disgust at the pigs they kept. The taboo probably was reinforced by the disease trichinosis, which could be contracted from eating insufficiently cooked pork.

The European wild boar (above) has been introduced to countries outside its native range, including the United States, and mixed with feral pigs (domestic pigs gone wild).

The earliest domestication date for pigs is uncertain, but it is unlikely to have been before the Neolithic period, with its agricultural revolution. There is evidence that wild boars were domesticated from local races, thus producing domestic pigs of various sizes. However, it is difficult to trace the lineage of present-day domestic pigs, because there also is evidence that in prehistoric times there was importation of breeds from one part of Eurasia to another, as well as some selective breeding, so there was a mixture before modern selective breeding began.

Pigs and people

Although the domestic pig has been bred mainly for its flesh and for its fat, that is far from the end of the history of its relationship with humans. Pig bristles have been used for making brushes, and pig hide has been used in making sandals and other leather goods. Pig bones may even be ground up for bonemeal fertilizer. The pig has also played a variety of roles as a living animal. There was a curious custom among Roman armies of swearing an oath on a pig or a piglet. Pigs have also been used for pulling carts and have been trained to detect truffles, which the pig's owner would then dig up. In southern India, a pig was once observed rounding up a herd of buffalo, like a well-trained sheepdog, and the ancient Egyptians used pigs for treading in corn, their sharp hooves making holes of the correct depth for the seed to germinate.

In medieval England, pigs were trained as pointers and retrievers by poachers. In the New Forest, southern England, for example, commoners were forbidden to keep any but the smallest dogs, those species capable of passing through King Rufus' Stirrup, an iron stirrup 10½ inches (27 cm) high by 7½ inches (19 cm) across, so they kept pigs as hunting companions instead.

WILDCAT

THE WILDCAT OF EUROPE and Asia, *Felis sylvestris*, resembles the domestic tabby but is more heavily built. The overall length is about 28–44 inches (70–110 cm), of which about 11 inches (28 cm) is tail, but there is a record from Scotland of one that reached 3¾ feet (1.13 m) total length, and there is an unconfirmed report of a wildcat shot in Scotland that measured 4¼ feet (1.28 m). The weight of a male averages 11 pounds (5 kg) but may exceed 17½ pounds (8 kg); one wildcat from the Carpathians weighed nearly 33 pounds (15 kg). The female is usually two-thirds of the male's weight.

The wildcat has a squarish head and a more robust body than the domestic cat. The thick, bushy tail appears to have a club shape; it bears three to five black rings, ending in a long, black tip. The limbs, too, are longer than those of the domestic cat. The fur is long, soft and thick and is mainly yellowish gray, but there is plenty of variation. Individuals differ in their dark brown markings: some have vertical stripes running down the flanks, whereas in others these are broken up to form spots. Since there seems to have been considerable crossbreeding with feral domestic cats, some of the variations in the pattern of the coat may be due to this. A few strong, dark longitudinal stripes occur over the brow.

In Europe the wildcat is now confined chiefly to the mountains, especially in the Balkans, and in Britain it is restricted mainly to the Scottish Highlands. It also extends into the Near East, with related species occurring in Central Asia. The subspecies *F. s. lybica* is found all over Africa, with the exception of the deserts and tropical forests.

Wild and inaccessible home

In Europe, the wildcat lives on wooded habitats, remote mountainsides, savanna or steppe. It hunts alone, prowling far and wide at night in search of prey, with peaks of activity at dawn and at dusk. In the fall there is a tendency to hunt by day, although for most of the year the wildcat hides during the hours of daylight. Like the domestic cat, it shelters from the rain and loves to bask in the sun on a bough or a rock. Dens range from a mere cleft between rocks to a vacated fox earth, badger set or hollow among tangled tree roots.

The home range covers an area of 500 acres (2 sq km), although the wildcat may wander much farther in search of prey in winter or in the breeding season. Within the range a wildcat maintains use of several dens. The wildcat marks its range by spraying urine on prominent landmarks or by depositing dung on grass tussocks

The ancestor of today's domestic cats is the African wildcat,* Felis silvestris lybica. *Another close, possibly conspecific relative is the Chinese desert cat,* F. bieti, *of East Asia.

The true European wildcat (above) closely resembles its domestic relative and feral cats (domestic cats turned wild) can also be hard to identify. So secretive is the wildcat that sightings are scarce.

or mounds. It may also rely on trunk-scratching to leave olfactory and visual markers. There may be overlap between the ranges of two neighbors, though seldom between cats of the same sex, and an exclusive core area is defended within every range. When wildcats meet, younger animals are subservient to their seniors.

The voice of the wildcat ranges from a miaow, an angry growl or a pleased purr to the typical small-cat scream or caterwaul on occasion. One naturalist writing in 1845 noted: "I have heard their wild and unearthly cry echo far into the quiet night as they answer and call to each other. I do not know a more harsh and unpleasant cry than that of the wildcat."

Poultry and lambs in peril

The wildcat's main food is small rodents, birds, and rabbits, but it eats any small mammal or bird it can catch, as well as fish and insects. It captures small rodents either by sitting in wait or by foraging. To tackle larger prey, such as a rabbit, the wildcat stalks and then rushes it, delivering a powerful bite to the back of the neck to sever the spinal cord. It also sits in wait above rabbit holes. The wildcat can create havoc among poultry or lambs, this being the main reason it has been

WILDCAT

CLASS **Mammalia**

ORDER **Carnivora**

FAMILY **Felidae**

GENUS AND SPECIES ***Felis sylvestris***

WEIGHT
6½–17½ lb. (3–8 kg)

LENGTH
Head and body: 20–30 in. (50–75 cm); tail: 8½–14 in. (21–35 cm)

DISTINCTIVE FEATURES
Similar in appearance to domestic cat, but larger and more thickset, with broad head, long fur and bushy, club-shaped tail

DIET
Small mammals, birds, reptiles, insects

BREEDING
Age at first breeding: 1 year; breeding season: often January–March, but can be year-round; gestation period: 65 days; number of young: 2 to 6; breeding interval: 1 year, but occasionally 2 litters per year

LIFE SPAN
Up to 15 years in captivity

HABITAT
Various, including forest, woodland, rocky country, steppe, savanna, mountain

DISTRIBUTION
From Scotland, Portugal and Spain east across southern and central Europe into western China and India; also found throughout Africa, excluding deserts and tropical forests

STATUS
Range and size are in decline, due mainly to human persecution, lack of prey and interbreeding with domestic cat

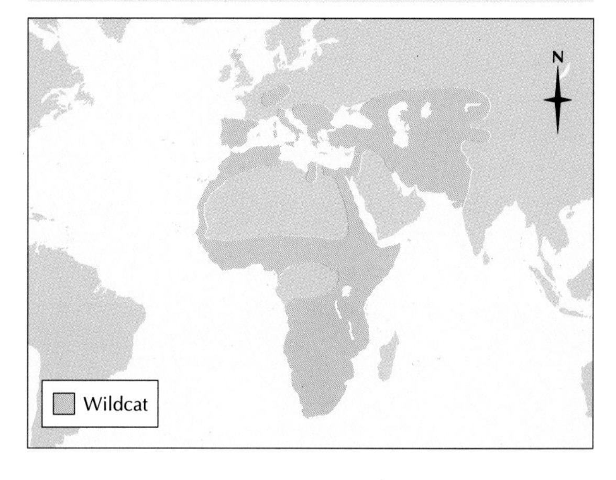

eradicated from much of its former range. Like a domestic cat, a wildcat may occasionally eat grass, probably to aid digestion.

Fierce mother

Wildcats usually breed between January and March, a litter of two to six kittens being born in April or May. The gestation period is 65 days. The kittens are born in a nest made by the female in some remote rock cleft or hollow tree, well away from the male, who may kill his own young. The mother rakes in a little grass or bracken to line the nest. The kittens weigh 3½–5¾ ounces (100–163 g) at birth and their fur is a pale ground color with grayish brown tabby markings. The female is very fierce while she has her kittens in the nest and attacks any animal, no matter what its size, that dares to intrude. Even the kittens spit and fight if handled. Very few have been taken alive, and all are said to have proved untamable.

The kittens open their eyes, initially blue, at 10–13 days. They leave the nest at 4–5 weeks old, but do not go hunting with the mother until they are 10–12 weeks old, and are not weaned until they are 4 months old. They leave the mother to fend for themselves when 5 months old, by which time their eyes have changed to the adult amber color. For a while the young are nomadic, until they establish their own home range. Males may remain footloose until well into adulthood, whereas females soon become established somewhere in preparation for breeding, when they will need a good-sized hunting range.

Persecuted by humans

The wildcat now has few natural predators, because most of the large predators in its range have been wiped out. Large eagles, such as golden eagles in Scotland, may take kittens. Nevertheless, humans remain a danger because of the threat the wildcat poses to poultry and lambs. Many years ago, when it was widespread over Europe, the wildcat was hunted for sport. It is one of the fiercest members of the cat family and its resilience, when tested, is remarkable.

The wildcat's decline began centuries ago, with the wholesale clearance of woodlands. In Britain, for example, the bone record reveals that the species roamed as far south as Berkshire in southeast England, but from medieval times on it gradually retreated north in the face of human expansion and persecution. By the turn of the 20th century its numbers were at an all-time low. However, the outbreak of war in 1914 resulted in British attention being focused abroad and the pressure was lifted from the wildcat in its Scottish stronghold. The expansion of conifer forestry during the century provided more habitat both for the cat and for many of its prey species.

A female wildcat with kittens in the Bavarian Forest National Park, southern Germany. Females become receptive to mates in winter and usually give birth in April or May.

WILDEBEEST

A male blue wildebeest grazing. The long tail is used to make fly whisks, which are a symbol of rank in eastern Africa and southern Africa. Both sexes have horns.

THESE COWLIKE ANTELOPES are known as gnus or, more frequently, by their Afrikaans name, wildebeest. They are up to 5 feet (1.5 m) at the shoulder, with short, thick necks and large heads with a tuft of long hair on the muzzle, a throat fringe and a mane. The male weighs up to 600 pounds (275 kg), the female about 20 percent less.

There are two species of wildebeest, the black wildebeest, *Connochaetes gnou,* also known as the white-tailed gnu, and the blue wildebeest, *C. taurinus*, or brindled gnu. The former has a long-haired white tail and forward-curving horns. The latter, slightly the larger of the two, has an equally horselike tail, but darker and laterally curved horns. Whereas the black wildebeest is dark brown to black in color, the blue wildebeest is silver gray with brownish bands on the neck, shoulders and the front part of each flank. The black wildebeest is restricted to southern Africa, while the blue wildebeest occurs in both southern Africa and east Africa.

Following the young grass

Blue wildebeest move about their range with the seasons. In the wet season they are scattered over the plains, then in the dry season they move along streams through the surrounding bush, seeking new grass produced by local showers. In the dry season the movements between these limited areas of new grass lead to massing, and huge numbers are visible in one place, the animals covering as much as 30 miles (50 km) in a day. The wildebeest feed in the morning and evening, seeking shade in the heat of the day. They tend to follow anything moving in a determined manner, including other wildebeest, other animals and even vehicles. Wildebeest feed on several species of grass, but favor only the fresh young growth, when the sprouts are not more than 4 inches (10 cm) high. The practice of regularly burning grass is beneficial to wildebeest.

Predation, then disease

The rut (mating season) often takes place during April and May. The peak of calving comes 9 months later, in January and February, when the plains are green. The calves are well developed and can follow their mothers within 4–5 minutes of birth. There is, however, a heavy loss through predation and through calves becoming separated from their mothers. Within a few weeks almost half the season's crop of calves are dead.

When the surviving calves are 6–7 months old, a rinderpest epidemic further reduces their numbers, and this lasts until 11–12 months after their birth. This outbreak is known as yearling disease. It arises at this time as a result of the calves' loss of their initial colostral immunity and does not affect adults. The colostral immunity is imparted by the mother to the calf in the first flow of her milk. This milk is called colostrum and contains proteins that pass directly through the calf's stomach wall into the bloodstream. These proteins include some of the mother's antibodies, which temporarily protect the infant wildebeest from disease.

The severity of yearling disease depends on the density of the population, but there is an average loss of 80 percent of the calves each year. The 20 percent left form 8 percent of the population. Because the numbers of wildebeest remain steady from year to year, this means that 8 percent of adults are lost each year. A calf stays with its mother until the next one is born, after which the cow prevents the elder calf from suckling. Bull yearlings then form separate herds, while the young cows remain in the cows' herd.

The rut takes place during the dry season. When the moving herds halt where food is abundant, some of the males establish harems, which

WILDEBEEST

CLASS **Mammalia**

ORDER **Artiodactyla**

FAMILY **Bovidae**

GENUS AND SPECIES **Black wildebeest, *Connochaetes gnou*; blue wildebeest, *C. taurinus***

ALTERNATIVE NAMES
White-tailed wildebeest, white-tailed gnu (*C. gnou*); white-bearded wildebeest, brindled gnu (*C. taurinus*)

WEIGHT
255–600 lb. (115–275 kg)

LENGTH
Head and body: 5–8¼ ft. (1.5–2.5 m); shoulder height: 3¼–5 ft. (1–1.5 m); tail: 14–24 in. (35–60 cm)

DISTINCTIVE FEATURES
Black wildebeest: dark brown/black; black mane; white tail. Blue wildebeest: gray or silver; mane and tail darker. Both species high at shoulder with large horns.

DIET
Plant matter

BREEDING
Age at first breeding: 2–3 years (female); breeding season: varies according to locality but births usually at start of rainy season; number of young: 1; gestation period: 8–9 months; breeding interval: 1 year

LIFE SPAN
Up to 20 years in captivity

HABITAT
Open, grassy plains

DISTRIBUTION
Black wildebeest: southern Africa; blue wildebeest: east and southern Africa

STATUS
Both species conservation dependent

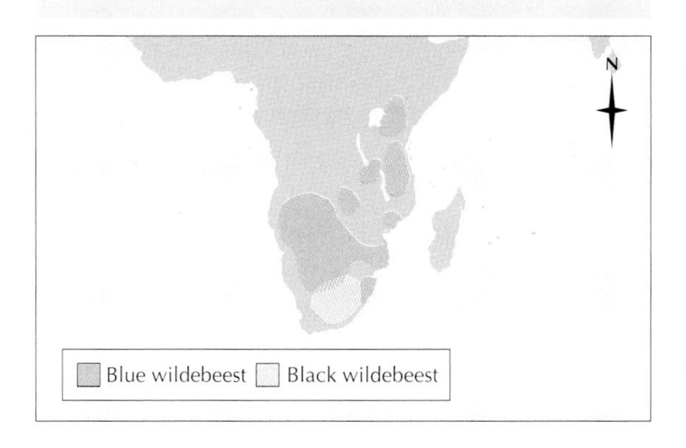

Blue wildebeest migrate en masse across the Mara, the northern extension of the Serengeti. There are an estimated 1.5 million wildebeest in the Serengeti region.

may contain as few as two to three females or as many as 150 females and young. Each male herds his harem tightly, running around and around it. Often, as many as three bulls herd the same harem. They are not aggressive toward one another, and they have no hierarchy. A herding bull runs with a rocking gait, head held high. The females, the young and the nonactive bulls keep the normal head-down position.

The harem forms the center of a temporary territory established by each bull or party of bulls. If a male approaches in rut position from the neighboring group, the bulls of the threatened group go to meet him. The nearest one rushes forward, the rivals drop to their knees, butt and spar, get up, snort, look around and then retire or repeat the procedure. The breeding groups are stable for several days, until the food is gone and the gnus must move on. Then the males retire to the margins of the migrating herd, either singly or in small groups. Each time there is a pause in the mass movement, the males round up harems and each time that the herd begins to move again they merge back into the mass. Only a small proportion of bulls form

Each year, in April or May, blue wildebeest migrate west out of the southern Serengeti and north into the Mara before returning to the southern Serengeti in November.

harems each time, so different individuals are active at each stop. Nonactive bulls can feed close to the breeding groups without being attacked, although if they graze too close they may be rounded up by other males along with the cows and calves. Wildebeest may live for 20 years. Each year, 83 percent of yearling cows breed, as well as 95 percent of the remaining cows. At birth there are 170 males to 100 females. After a year the ratio is 117:100, and after 1½ years the adult ratio of 108:100 is reached.

Lions most feared

By far the most important predator of wildebeest is the lion. An estimated 37 percent of blue wildebeest deaths are caused by predation, and lions are responsible for almost all of these. At least half of the lion's prey on the plains of the Serengeti, located in Tanzania and just over the border in Kenya, consists of wildebeest. In a study carried out in 1963 by Lee and Martha Talbot, 91.1 percent of wildebeest predation was by lions. The remainder was spread across a range of other predators: 3.3 percent was by cheetahs, 2.2 percent by leopards, 2.2 percent by spotted hyenas and 1.1 percent by wild dogs. Wildebeest defend themselves against cheetahs and wild dogs by forming a circle, in the manner of musk-oxen (discussed elsewhere). Poaching and accidents also take a toll on wildebeest populations, the number of deaths attributable to these causes being about twice that caused by predators other than the lion.

Sharing out the food

The ecology of the hooved animals on the Serengeti provides a striking lesson in natural land use. The wildebeest feed only on young shoots, as discussed earlier. Zebra feed on the same grasses but at a later stage of growth. Topi, *Damaliscus lunatus*, also feed on the same plants but on old grass. Further animal species feed on different grasses growing in the same places. Thus, the Serengeti can support a far greater quantity of wild hooved animals than domestic cattle and is potentially much more valuable as a source of protein to wild animals.

The enormous wastage of wildebeest calves in their first year gives great scope to natural selection and may surprise those accustomed to thinking of large hooved animals as slow to breed and consequently slow to evolve. Several races of blue wildebeest exist. To support wildebeest a habitat must provide new grass year-round. Such places are rare, so the wildebeest tend to occur in areas that are isolated from each other. Differences arise between the various groups just as they do in animals separated from others of their kind through living on islands (see Darwin's finches, discussed elsewhere).

WILD GOAT

Truly wild goats, **C. aegagrus,** *may first have been domesticated more than 10,000 years ago. The domesticated goats,* **C. hircus,** *above, were photographed on the Mull of Kintyre, Scotland.*

THERE ARE EIGHT SPECIES of goat: the wild goat, *Capra aegagrus*; the domestic goat, *C. hircus*; the Alpine ibex, *C. ibex*; the walia ibex, *C. walie*; the West Caucasus tur, *C. caucasica*; the East Caucasus tur, *C. cylindricornis*; the Spanish ibex, *C. pyrenaica* and the markhor, *C. falconeri*. Truly wild goats are found in Turkey, Iran, Turkmenistan, Pakistan, Central Asia and on a number of Greek islands, including Crete.

Goats are 3–5⅘ feet (1–1.6 m) long in body and head. The tail is 6–8 inches (15–20 cm) long, they are 28–40 inches (70–100 cm) tall at the shoulders and weigh up to 55–209 pounds (25–95 kg), the males being larger than the females. The horns of males are sweeping and scimitar-shaped, grow up to 52 inches (130 cm) long and are compressed sideways and ornamented along the inner front edge with large knobs. The horns of the females are shorter and more slender than those of the males. The coat is typically reddish brown in summer, grayish brown in winter, with black markings on the body and limbs. However, there is considerable variation in coat color, according to species.

Desert-making goats

Goats usually live in rugged, rocky or mountainous country, but are sometimes found on lowland plains. Where they are hunted, they become extremely wary and difficult to stalk, because their sure-footed skill as they progress from rock to rock makes them hard to follow. They generally move about in herds of 5 to 20, led by an old female. When living on mountains, they may go up almost to the snowline, but in winter they migrate to lower levels, returning in spring to the fresh pastures. Goats do not sleep; they merely have periods of drowsiness.

Goats are able to live in fairly harsh environmental conditions that would deter other species. They are able to survive by eating desert scrub, bushes, herbs and trees, insufficient nourishment for other grazers and browsers, such as sheep and cattle. They eat straw and have been seen to scratch their backs with straws held in the mouth. Like sheep, goats chew the cud, but whereas sheep take mainly grass, goats browse chiefly on leaves and twigs as well. They eat desert scrub and are even able to climb into trees

Wild goats, such as this Cretan wild goat, C. a. cretica, *are distinguishable by their distinctive scimitar-shaped horns.*

to browse. Goats have been seen to jump onto the backs of animals such as donkeys to reach the lower boughs, and from there move to higher and higher boughs by jumps. They readily take bark and paper and are notorious for eating linen cloth. In this they are helped by protozoa living in the gut that predigest cellulose. Domestic goats eat the foliage of yew, which may be fatal to horses and cattle, and suffer only a temporary diarrhea. Released on oceanic islands, goats have reduced earthly paradises to barren soil with only low vegetation. In western Asia and the Middle East, herds of goats have contributed to the formation of deserts.

Climbers from birth

A female goat more than 2 years old is known as a nanny goat, and the male is a buck or a billy goat. Both are relatively recent names, the first having been used since 1788 and the second since 1861. Mating normally takes place in the fall in temperate zones, usually from August to February. In areas nearer to the equator, however, the females come into estrus (breeding condition) year-round, creating the possibility of more than one litter in a year. The offspring, known as kids, are born 150–170 days later in domestic breeds, and are able to run shortly after birth and soon become adept at climbing. There are one, sometimes two young at a birth, exceptionally three or four. Sexual maturity is reached in about 12 months, at which time the male is known as a buckling and the female as a goatling. Goats live for up to about 20 years.

WILD GOAT

CLASS **Mammalia**

ORDER **Artiodactyla**

FAMILY **Bovidae**

GENUS AND SPECIES ***Capra aegagrus***

ALTERNATIVE NAMES
Bezoar (*C. aegagrus*); Sind ibex, (*C. a. blythi*); Chiltan goat, (*C. a. chiltanensis*); Cretan goat (*C. a. cretica*); domesticated goat (*C. hircus*)

WEIGHT
55–209 lb. (25–95 kg)

LENGTH
(*C. hircus*) Head and body: 3–5⅗ ft. (1–1.6 m); shoulder height: 28–40 in. (70–100 cm); tail: 6–8 in. (15–20 cm)

DISTINCTIVE FEATURES
Stocky build; broad hooves; beard; horns usually curve backward, away from head; considerable variation in coat color, from pure white to red-brown or black

DIET
Grass; leaves and herbage

BREEDING
(*C. hircus*) Age at first breeding: about 3 years; breeding season: varies according to location; number of young: 1 (2 or 3); gestation period: 150–170 days; breeding interval: 1 year

LIFE SPAN
Up to about 20 years in captivity

HABITAT
Very varied

DISTRIBUTION
Afghanistan; Armenia; Azerbaijan; Georgia; Greece; Iran; Oman; Pakistan; Russia; Turkey; Turkemenistan

STATUS
(*C. hircus*) Estimated population 600 million

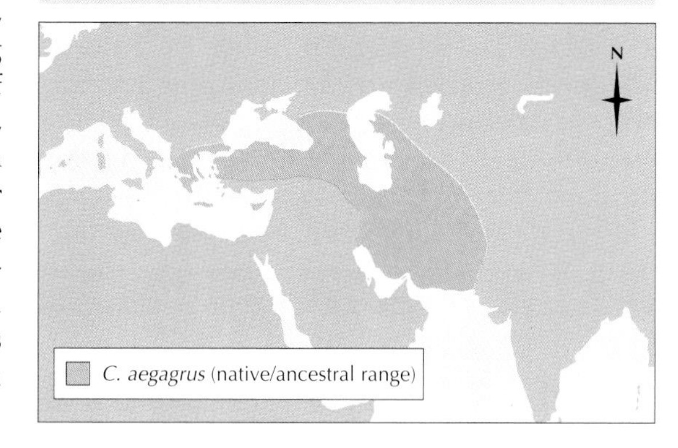

Indestructible ravagers

Goats probably have always been of more value for the milk that may be obtained from them than for their hair or flesh. More than 4½ million tons (4.1 million tonnes) of goat milk and 1⅕ million tons (1.1 million tonnes) of goat meat are produced every year. Goat flesh is generally regarded as somewhat rank in taste and the hair is short, but is sometimes used for spinning, especially that obtained from longer-haired goats such as the Turkish Angora, the Asian cashmere and the Russian Don. Leather is also produced from goat skin in some countries.

In the past, ships took goats on board to provide fresh milk as well as meat at sea. Ships' captains put goats ashore on oceanic islands for the use of castaways or to get rid of surplus. The marooned goats multiplied and, as on St. Helena and other islands, denuded the flora. In 1773, Captain James Cook put goats ashore in New Zealand. These subsequently became feral and multiplied. Later, goats were taken there for other purposes, to feed workers who were building roads and railroads for use in miners' camps, and to prevent introduced bramble, gorse and bracken from running amok.

In fact, the goats stripped the tree bark, ate shrubs, brought many native plants to the verge of extinction and cleared the ground of mosses that not only held water but also protected the topsoil from wind erosion. Their hooves cut the turf so that it was washed away by rain, adding to erosion problems. The natural home of a goat is barren hillside, and wherever goats go they convert the landscape into their natural habitat.

The speed at which goats multiply is one of their most distinctive features. In 1698, an English ship put into the harbor of Bonavista, Newfoundland. Two local people offered the captain all the goats he cared to take away. There were only 12 people living on the island and the goats were not only eating everything but also were so tame nobody could move about on the island without being followed by a group of them.

An historical note

In his *Chronicles of England*, first published in 1577, the English historian Raphael Holinshed noted: "Goats we have in plenty, and of sundry colors, in the west parts of England; especially in and toward Wales, and among the rocky hills, by whom the owners do reap no small advantage." Although Holinshed did not specify what the advantage to the farmers was, contemporary historians believe that goats were deliberately allowed to go wild in these regions by the local sheep farmers. In the Welsh mountains, grass grows lush in inaccessible places. Sheep attracted by the grass find that they cannot get down again and have to be retrieved. Wild goats, which are more skillful climbers than the sheep, climb the high rocks and eat the grass, thereby removing the temptation for the sheep, and have no difficulty in descending again.

Goats are native to barren, rocky landscapes and are capable of reducing areas of lush vegetation to a similar denuded state within a relatively short time.

WOBBEGONG

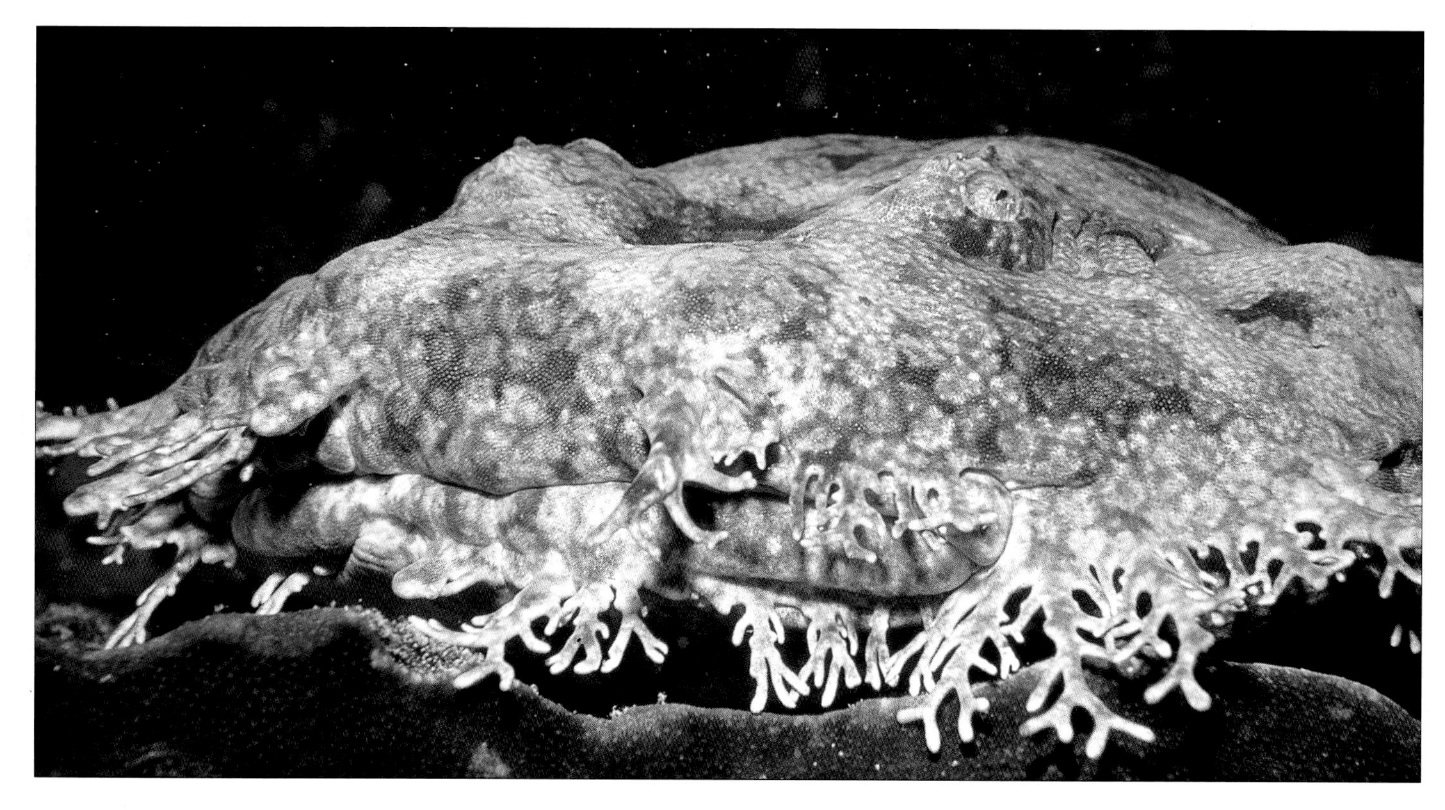

A tasseled wobbegong on Australia's Great Barrier Reef. Like other species, its head is fringed by a beard of weedlike flaps of skin.

THE WOBBEGONGS, OR CARPET sharks, are unusual among sharks in that they use ambush instead of speed to obtain food. Resting on the sea bottom, they look like rocks overgrown with seaweed, a perfect camouflage that enables them to pounce on unsuspecting victims. Most wobbegongs are small, but some grow to a length of 6–8 feet (1.8–2.4 m). The largest is the spotted wobbegong, *Orectolobus maculatus*, of Australia, which ranges from Queensland to Tasmania and grows up to 10½ feet (3.2 m).

A wobbegong is quite unlike a typical shark in shape. It has a stout, thickset, flattened body with a very broad head and a blunt, rounded snout with a wide, straight mouth. Its teeth are slender and pointed, those in the center of the jaws being the largest. Its eyes are small with folds of skin below them, and the wide, oblique slits of the spiracles are situated behind the eyes and lower down on the head. The last three or four external gill clefts on each side open above the bases of the pectoral fins, which are broad and sometimes rounded. The two dorsal fins are comparatively small and set well back. The anal fin either extends to, or is actually joined at its base to, the lower lobe of the tail fin, which is long and asymmetrical with a notched tip. The skin is covered with small, rough denticles (toothlike scales). The skin color varies from one species to the next, or even among individuals, but the base color is usually brown, yellowish or gray with distinctive mottled or striped markings. What distinguishes the wobbegong from any other shark is the fringe of fleshy lobes or flaps of skin around the sides of the head and mouth, which resemble fronds of seaweed when the shark is at rest. There are six species of wobbegongs living in the seas around China, Japan, Indonesia and Australia.

Sluggish existence

A wobbegong's heavy body is not built for speed like most other sharks, which hunt actively, but its camouflage gives it an equally effective means of obtaining food. It spends most of its time lying hidden on the sea bottom among the rocks and weeds, only coming to life when a fish passes by for it to snap up. It is further concealed by waving its flaps of skin so that they look even more like seaweed. A wobbegong does not need to keep moving, as do many typical sharks, in order to breathe. It draws water into the gill chamber through the spiracles, in much the same way that skates and rays do.

United by scents

Male wobbegongs are attracted to females by pheromones (sexual scents) released into the water during the breeding season. Like the nurse sharks, the wobbegong is ovoviviparous (the

WOBBEGONGS

CLASS **Elasmobranchii**

ORDER **Orectolobiformes**

FAMILY **Orectolobidae**

GENUS AND SPECIES **Spotted wobbegong, *Orectolobus maculatus* (detailed below); ornate wobbegong, *O. ornatus*; northern wobbegong, *O. wardi*; Japanese wobbegong, *O. japonicus*; cobbler wobbegong, *Sutorectus tentaculatus*; tasseled wobbegong, *Eucrossorhinus dasypogon***

LENGTH
Up to 10½ feet (3.2 m)

DISTINCTIVE FEATURES
Flat-bodied, sluggish shark; dark back with pale O-shaped markings obscuring darker saddles; tail fin with short upper lobe, strong terminal lobe and no ventral lobe

DIET
Bottom invertebrates, bony fishes

BREEDING
Male attracted to female by her scent; female ovoviviparous (gives birth to hatched young); number of young: up to 37, usually about 20; size at birth: about 8½ in. (21 cm)

LIFE SPAN
Not known

HABITAT
Found on continental shelf, from intertidal zone down to at least 360 feet (110 m); common on coral and rocky reefs, under piers and on sandy beds

DISTRIBUTION
Indian Ocean and western Pacific; unconfirmed reports from Japan and South China Sea

STATUS
Not known

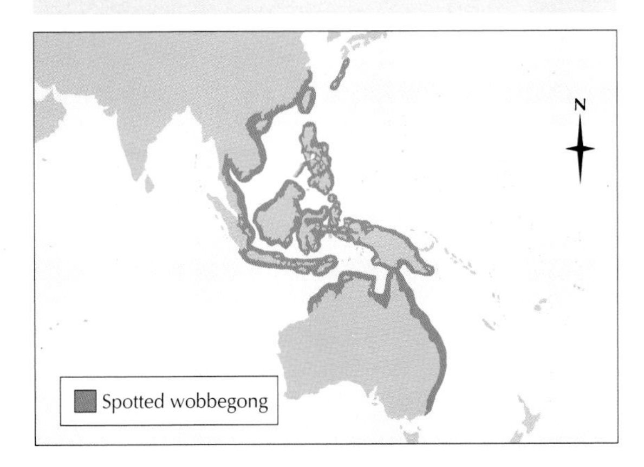

A subtly reticulated skin camouflages the wobbegong against the rocky, sandy or weedy bottom on which it lies in wait for fish and invertebrate prey.

young hatch from the eggs before they leave the female's body) and produces large litters, typically of about 20 young.

Danger to humans

Wobbegongs are not in the first league of sharks known to be a danger to humans, but they frequently attack people wading in shallow water and are treated with caution by divers. Their camouflage makes them very difficult to spot on the seabed, and they react aggressively when trodden on. Their long, pointed teeth are quite capable of biting off a human foot.

A record from New South Wales, Australia, tells of a wobbegong attacking a spear-fisher at work in Shellharbour in 1953. The fisher was using a snorkel and wearing an underwater mask with a bright metal band. He was trying to spear a dying grouper and, running out of shafts, he surfaced and borrowed another gun. As he shot a spear into the fish, it came out of its cave followed by a large, brown wobbegong shark with three or four tentacles hanging from its lip. The spear-fisher swam quickly for about 15 yards (13.7 m), but the shark rushed at him in attack. It seemed likely that the wobbegong was going for the mask with its metal band, since bright objects and metal are known to attract sharks. It tore away the face piece and snapped off the snorkel. The rush of the attack was so dramatic that both the shark and the fisher were hurled from the water. This enabled the spear-fisher to escape with injuries only to the underside of his chin and to his face and nose.

The wobbegong has little or no value as food. However, its colorful, variegated skin makes it valuable in the shark leather industry.

WOLF SPIDER

A wolf spider, genus* Lycosa*, attaching an egg sac to her abdomen. She carries the sac with her wherever she goes.

Wolf spiders, like the carnivores after which they are named, run their prey down instead of making a silken snare like so many other spiders. These are small to medium-sized spiders, the largest species having a body less than 1 inch (25 mm) long, with legs of the same length. They are dark or drab in color, the body and legs being covered with short bristles. Wolf spiders have strong jaws. Their most characteristic feature is the four pairs of eyes: two large eyes facing forward, two facing sideways and four smaller eyes that also face forward. As in other spiders, the eyes are simple.

Wolf spiders are widespread across all the continents except Antarctica. The total number of species is around 3,000 in the family Lycosidae and 600 in the Pisauridae. There are about 125 species of wolf spiders in North America. The 50 species in Europe include the true tarantulas of the genus *Lycosa*.

Wolflike in many ways

Wolf spiders are often numerous, especially among leaf litter. They tend to be more active at night or under overcast conditions, but they can sometimes be seen in large numbers by day, running over dead leaves. They shelter by day in small burrows dug in soft soil. Some species line the burrow with silk, whereas others also have a silken tube running out a short way from the mouth of the burrow. The silk is never used as a snare but more as an ambush. As with most spiders, lycosid wolf spiders prey on small insects, which they pursue and grab with their strong jaws. They then chew the victim to a pulp and suck the juices through a very small mouth, too small to admit any but the smallest particles. Pisaurid wolf spiders, some of which are found in wetlands, include fish fry and tadpoles in their diet.

Wooing by waving

Wolf spiders have relatively keen sight, and among the lycosids courtship is conducted by visual signals. A male ready to breed sets out in search of a female. He stations himself in front of her and waves his long pedipalps up and down in the manner of semaphore arms. The pedipalps (appendages near the mouth) are usually black and conspicuous against the drab body. In some species, parts of the front pair of legs are also black, and these are also held up and waved. In many species the male vibrates these legs; in others the male may vibrate the legs and abdomen. If the female is not receptive at first, she may respond later, after witnessing the male's display. During mating, the male places his sperm, previously deposited on his pedipalps, into the female's genital pore.

The female lycosid lays her eggs in a spherical or lens-shaped silken egg sac spun especially for the purpose. She attaches this egg sac to the rear end of her abdomen, carrying it around with her wherever she goes and devoting a high degree of care to it. She seeks shelter when it is raining to preserve the egg sac, and should it get wet she will, at the earliest opportunity, tilt her head and body down to hold the cocoon up to the sun and dry it (though in arid habitats some lycosids deliberately moisten the egg sac). If the egg sac becomes detached, the female retrieves it, reattaching it to her abdomen. In tests, if the sac is taken away and a small white object, such as a piece of paper, is placed on the ground near the spider, she retrieves this instead.

Family transport

When the eggs hatch, the spiderlings remain for a short while in the egg sac until it splits, under favorable weather conditions. They then climb onto the mother's back, and she transports them. In some species the brood is so numerous that the spiderlings cover the mother's back several layers deep. The spiderlings appear not to feed during this time. Should one fall off, the mother does not halt for it to regain its position on her back or to help it up. The spiderling must either quickly climb up one of her legs to her back or be left to fend for itself. The spiderling may pay out a line of silk as it tumbles, using this as a safety line with which to reel itself back on board.

Nursery webs

Female wolf spiders of the family Pisauridae show a different method of parental care. Shortly before the eggs are due to hatch in June or July, a female attaches the sac to vegetation and then weaves a silken tent over the mass. When the spiderlings hatch, they remain huddled within the web, the female straddling the entire structure to guard them from harm. The spiderlings undergo their first two molts within the safe haven and then disperse. For this reason, species of the genus *Pisaura* have come to be known as nursery web spiders.

WOLF SPIDERS

PHYLUM **Arthropoda**

CLASS **Arachnida**

ORDER **Araneae**

FAMILIES **Lycosidae, Pisauridae**

GENUS AND SPECIES **About 3,600 species in several genera, including *Pisaura, Lycosa* and *Pardosa***

LENGTH
Abdomen: ¼–1½ in. (5–40 mm)

DISTINCTIVE FEATURES
Generally brown or gray in color; long front legs covered with strong spines; four pairs of eyes, with two large eyes facing forward, two sideways and four smaller eyes

DIET
Invertebrates; fish fry; tadpoles

BREEDING
Male lycosid courts female by waving palps; mated female lays eggs in silken sac and carries it. Male pisaurid woos female with food gift; female carries egg sac, spinning nursery tent over it before spiderlings hatch

LIFE SPAN
Usually one season; rarely 18 months

HABITAT
Various, including wetlands and mountains

DISTRIBUTION
Worldwide, excluding Antarctica

STATUS
Common

Spider killers

The drab coloring of wolf spiders, their mainly nocturnal habits and the use of burrows by some species gives them a fair degree of immunity from attack. However, they do have two important predators. The first is the mantis fly, which preys on the eggs while they are still in the sac. Another danger comes from hunting wasps, which paralyze the spiders and use them to feed their larvae. These wasps fly low over the ground on a zigzag course in search of prey. When a wasp reaches the burrow of a wolf spider, it lands and starts to dig. Once the wasp is inside the burrow, the spider is easy prey.

The British arachnologist W. S. Bristowe has described touching a wolf spider *Arctosa perita* sitting at the mouth of its burrow with the tip of a fine grass stem. The spider grabbed the silk on one side of the opening and pulled it across, like a curtain, until only a small slit was left, which it then closed by spinning silk over it. A wolf spider may employ a similar defense technique to keep out a hunting wasp.

Mother's clear vision

The eyes are more important to wolf spiders than they are to web-spinning spiders. When a female is loaded with a brood of spiderlings, it seems at first that none of them ever obstructs her eyes. Closer study reveals that every now and then one or more spiderlings are pressed forward by the mob, almost blindfolding the mother. The moment this happens, she passes a pedipalp over her head and casually pushes the erring spiderlings away from her eyes.

A female* Lycosa *spider carries her brood of spiderlings. The spiderlings stay with their mother until their second molt, after which they disperse.

WOLVERINE

The wolverine has a number of adaptations to the cold, including furred feet to provide insulation from the snow and a dense coat that can shed ice crystals.

THE WOLVERINE, OR glutton, is the largest member of the weasel family (Mustelidae). However, it does not truly resemble a weasel, marten or stoat, being closer in appearance to a bear or badger. A full-grown male wolverine may be 4 feet (1.2 m) long, including almost 9 inches (23 cm) of tail. It can reach almost 18 inches (46 cm) at the shoulder and weigh up to 66 pounds (30 kg). The females are smaller and lighter. The wolverine is powerfully built and thick-bodied. Its legs are set wide apart and end in broad, powerful paws armed with long, sharp claws. The shaggy coat of dense fur is very dark brown above, with a pale brown band on the sides, and dark brown below. The wolverine ranges across the Arctic and subarctic regions of Europe and Asia and is found in North America from the Arctic south to the northern United States and California.

Wolverines live in cold evergreen forests up to 13,000 feet (4,000 m) above sea level. They are solitary except during the breeding season, inhabiting very large but definite territories. They hunt mainly at night, although at times when there are 24 hours of daylight, such as during the Arctic summer, wolverines tend to have a 3- or 4-hourly rhythm of alternate activity and rest. They do not dig burrows or make any permanent home, but they may adapt and line a burrow or use whatever other shelter is available where they are hunting. For example, they may tear down the timbers of a cabin to effect an entry and also wreck the contents, consuming what food they need and carrying away other articles. This has given wolverines an unwarranted reputation for destructiveness.

Ferocious opponent

The wolverine has an awkward gait but is a fairly fast runner. Unlike the smaller members of its family, it has little skill in stalking. At most, it may hide behind a rock in a kind of ambush, or drop from a branch of a tree onto an animal's back. It may also kill any caribou, *Rangifer tarandus*, that have become trapped in deep snow. Mainly, though, the wolverine depends for survival on its unusual ferocity in driving other predators from their food. The wolverine bares its teeth, raises the hair on its back, erects its bushy tail and emits a low growl. Even bears

WOLVERINE

CLASS **Mammalia**

ORDER **Carnivora**

FAMILY **Mustelidae**

GENUS AND SPECIES ***Gulo gulo***

ALTERNATIVE NAME
Glutton

WEIGHT
20–66 lb. (9–30 kg)

LENGTH
Head and body: 28½–42 in. (72–107 cm); shoulder height: 14–17¾ in. (35–45 cm); tail: 6¾–8¾ in. (17–23 cm)

DISTINCTIVE FEATURES
Coat dense, dark brown, yellow white on shoulders; sometimes white facial markings; feet furred

DIET
Small rodents, birds, frogs, plant matter, carrion; large ungulates caught in deep snow

BREEDING
Age at first breeding: 3 years; breeding season: April–August; number of young: 1 to 4 (born February–March); gestation period: 30–40 days (delayed implantation); breeding interval: 2 years

LIFE SPAN
20 years in captivity

HABITAT
Tundra, montane forest and rocky woodland

DISTRIBUTION
Scandinavia east to Siberia, Mongolia and China (Manchuria); Alaska, Canada and parts of northern United States; California

STATUS
Vulnerable; endangered or threatened in some parts of range

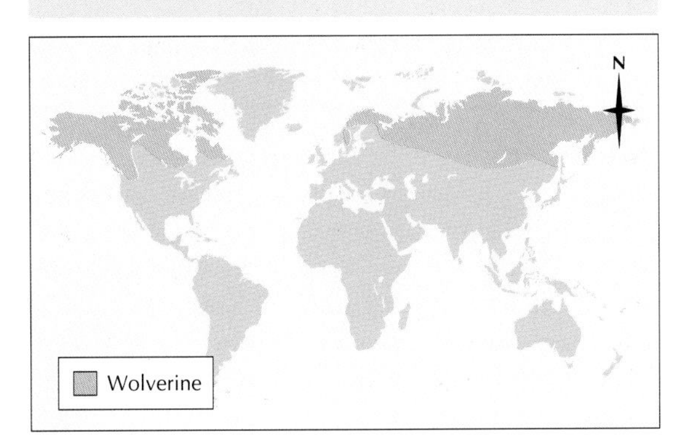

***A wolverine traveling through undergrowth in Finland. Two wolverine subspecies are generally recognized: the European wolverine,* Gulo gulo gulo, *and the North American wolverine,* G. g. luscus.**

have been driven from a carcass by such a display. The wolverine has remarkably strong teeth and jaws and is said to crunch large bones to powder and to snap branches up to 2 inches (5 cm) in diameter with ease. However, reliable reports suggest that, unless disturbed with young, it is not aggressive toward humans.

Trappers' tales

Among the exaggerated stories about the wolverine is its alleged skill not only in avoiding the snares of the fur trapper but also in robbing those traps in which martens and foxes have been caught. When it is satiated, it is said to exude its musk on any remaining carcasses to dissuade other animals from taking them. Tradition also holds that parent wolverines teach their young how to spring traps. The truth seems to be that the wolverine does sometimes rob traps, and there may have been occasions when a whole line of traps has been cleared, but as a rule the depredations are not as wholesale as most accounts suggest. As for its skill in avoiding traps, this seems to be explained by the small size of the traps used. Wolverines are usually caught by the toes in marten traps and often escape

Like a bear, the wolverine walks on the soles of its feet, the distribution of load enabling the animal to move over soft snow without sinking into it.

leaving a toe or two behind. There seems to be no evidence that they can avoid or escape the larger traps deliberately set for them.

Wide-ranging diet

The wolverine's diet is a wide-ranging one. Mice, rats, other small mammals, ground-nesting birds, ducks and even frogs are included. Above all, carrion, especially the kills of other carnivores, is eaten. The wolverine is reputed to be powerful enough to drag a carcass three times its own weight for some distance over rough ground. Uneaten food is cached, either covered with soil or snow or wedged in the fork of a tree. The wolverine has a reputation for eating more than any other carnivore, hence its alternative name of glutton, but probably the many stories about its excessive eating are also exaggerations.

Delayed implantation

The breeding season lasts from April to August, and the young, which usually number two or three, occasionally four, are born from February to March. In the past there was some uncertainty about the wolverine's gestation period, the records varying from 60–120 to even 183 days. Scientists now know that gestation lasts 30–40 days, implantation being delayed. The young are born in a hollow tree, among rocks or even in a snowdrift. They have thick, woolly fur at birth and are weaned at 8–10 weeks. They stay with their mother for up to 2 years, and then she drives them away to find their own territories and fend for themselves. Wolverines are sexually mature at 3 years of age and have been known to live in captivity for 20 years.

Persecuted by humans

Being so powerful, the wolverine has little to fear from natural predators. It has, though, been persecuted by people for its destructiveness and because of its reputation for killing caribou. For more than 100 years, attempts have been made in Norway to stamp out the wolverine, and premiums have been paid for each one destroyed. The Inuits hunt it for its fur because it does not hold moisture, which would freeze, and so is invaluable for trimming the hoods of their parkas. However, although its numbers have been reduced everywhere due to such persecution, the wolverine still is not uncommon in parts of its range.

WOMBAT

Like their closest relative, the koala, *Phascolarctos cinereus*, wombats look like little bears, but they are more like badgers in their habits, and they are often called badgers in Australia. Wombats can be up to 4 feet (1.2 m) long and weigh up to about 88 pounds (40 kg). They are thickset with little or no tail, their legs are short and strong and their toes are armed with stout claws used in digging. Wombats' teeth are unlike those of other marsupials, being more similar to those of rodents. The 24 teeth are rootless, and the upper and lower jaws each have two incisors, like those of a beaver. There are traces of cheek pouches.

There are three species of wombats, none of which overlap in their distribution. The common wombat, *Vombatus ursinus*, also called the coarse-haired wombat, lives in the hilly or mountainous coastal regions of southeastern Australia and on Tasmania and Flinders Island in the Bass Strait. It is the largest of the three species and has a naked muzzle, rounded ears and coarse fur ranging from a yellowish buff to dark brown or black.

The southern hairy-nosed wombat, *Lasiorhinus latifrons*, lives in the sandy or limestone coastal country and drier inland areas in the southern half of South Australia. Its sister species, the northern hairy-nosed wombat, *L. krefftii*, is Australia's most endangered mammal. Previously present also in New South Wales and Victoria, it is now found only in Queensland's Epping Forest National Park. Both species of hairy-nosed wombats can be distinguished from the common wombat by their slightly smaller size, larger skull bones and relatively pointed ears. They also have soft, silky fur and, as their name suggests, a hairy muzzle.

Shy and nocturnal

All three species of wombats are nocturnal and shy and therefore difficult to observe in the wild. They are the largest burrowing animals in the world and sleep by day in tunnels dug out with the powerful claws of their forefeet, the soil being thrust back with the hind feet. The burrows are large. They are usually 10–15 feet (3–4.5 m) long but can be as much as 100 feet (30 m) in length, and one series of burrows is reported to have been 880 yards (800 m) long. The entrance to a wombat burrow is large and arched, and a short distance away there usually is a shallow depression scraped in the surface of

There are three subspecies of common wombats: a mainland race (below), a Tasmanian race and a Flinders Island race. The common wombat is notable for its cube-shaped droppings.

Like all wombats, the southern hairy-nosed wombat is immensely strong and a speedy runner. Known as the bulldozer of the bush, it can move at 25 miles per hour (40 km/h).

the ground, at the base of a tree, where the wombat goes to bask in the morning sun. At the end of the burrow there is a sleeping chamber that contains a nest of bark. Wombats are mostly solitary, although occasionally they share burrows. To avoid competing for feeding areas, they scent-mark their territories and use aggressive vocalizations. When it is threatened, a wombat dives into its burrow, using its thick-skinned rump as protection from its attacker. If necessary, the wombat can crush an intruder against the burrow wall or ceiling. Although heavily built, the wombat is quick in its movements and can run swiftly for short distances. The only sound it makes is a hoarse grunting cough, rather like that made by a large kangaroo.

Diet of grass

There are often well-defined paths leading from a wombat's burrow to feeding areas in open country. Wombats eat mainly grass and roots and occasionally fungi, soft green moss and the inner bark of certain trees. They tear out and grasp the grass stems with their forefeet, and they sometimes damage pasture and crops near settled areas. Their diet has a high silica content that wears down their teeth, which in response continue to grow throughout the animals' lives. Wombats also carry bacteria in their gut to help them digest their food.

During May to July the female gives birth to one young, which is carried at first in the pouch, in the usual way of marsupials. Unlike those of most marsupials, however, wombat pouches open to the rear so that they do not get filled with soil as the animal burrows. After about 6–9 months the juvenile wombat runs free but stays

WOMBATS

CLASS **Mammalia**

ORDER **Diprotodontia**

FAMILY **Vombatidae**

GENUS AND SPECIES **Common wombat, *Vombatus ursinus* (detailed below); northern hairy-nosed wombat, *Lasiorhinus krefftii*; southern hairy-nosed wombat, *L. latifrons***

ALTERNATIVE NAME
Coarse-haired wombat (*V. ursinus* only)

WEIGHT
Up to 88 lb. (40 kg)

LENGTH
Head and body: 28–48 in. (70–120 cm); shoulder height: 14 in. (35 cm)

DISTINCTIVE FEATURES
Coat coarse, yellowish buff to black; stocky body; short, powerful limbs; rounded ears; stubby tail; pouch opening to rear

DIET
Grasses, herbs, roots

BREEDING
Age at first breeding: 2 years; breeding season: spring/summer; number of young: 1 (2); gestation period: about 28 days

LIFE SPAN
Up to 26 years in captivity

HABITAT
Flat, sandy semiarid grassland; eucalyptus or acacia woodland with patchy scrub and ground cover of native grasses

DISTRIBUTION
***V. ursinus*: southeastern Australia, Tasmania, Flinders Island; *L. krefftii*: Queensland; *L. latifrons*: South Australia**

STATUS
***V. ursinus*: vulnerable; *L. krefftii*: endangered; *L. latifrons*: uncommon**

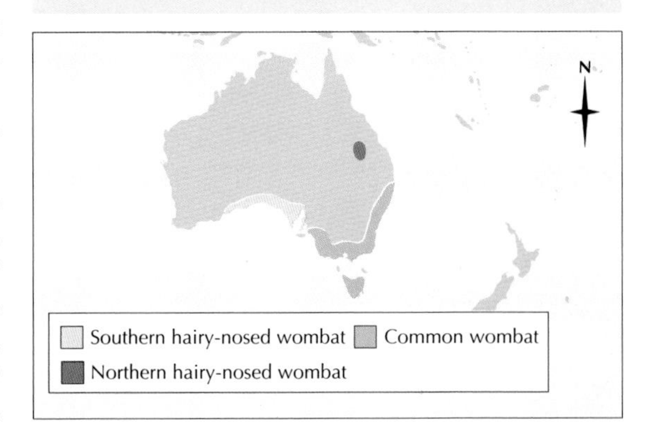

with the mother until the end of the year. During this time she feeds it on sword grass, pulling out the stems singly and dropping them on the ground so that the youngster is able to feed on the tender bases.

Humans the chief persecutors

Wombats have few natural predators. In all likelihood, only a dingo, *Canis dingo*, could take an adult, although a wedge-tailed eagle, *Aquila audax*, might possibly snatch a youngster caught out in the open by day. However, wombats have suffered severely at the hands of humans. In many areas wombats have disappeared entirely. Although wombat skin is not used commercially, the aboriginal peoples of Australia make string out of the fur of the hairy-nosed wombats, coiling it around their hair. Wombats have been ruthlessly banished from settlements from early days because of their habit of tearing down fences to reach the grass in sheep pastures or cultivated crops of various kinds. Besides their general appearance, another similarity with badgers is that wombats have been killed for their hams (cuts of meat taken from the thigh). Their burrows have been destroyed because of the danger they pose to horse riders and because rabbits sheltering in them could not easily be exterminated. Nowadays the wombats' biggest threats are from being struck by cars when crossing roads and from habitat loss due to agriculture. Dogs, overhunting and poisoned baits intended for rabbits have also taken their toll on wombat numbers. A major recovery program is currently underway, funded by the Queensland government, to reestablish the remaining population of the northern hairy-nosed wombat.

Historical account

The first account of the common wombat in New South Wales was supplied in the 18th century by a former convict, James Wilson, while on a journey into the territory's southern highlands across the Nepean River. His account was written down by one of his companions, a young servant to Governor John Hunter, who noted, when they were evidently near the present town of Bargo on January 26, 1798: "We saw several sorts of dung of different animals, one of which Wilson called Whom-batt, which is an animal about 20 inches [50 cm] high, with short legs and a thick body with a large head, round ears, and very small eyes, is very fat, and has much the appearance of a badger."

L. latifrons *is about 3 feet, (90 cm) long and may weigh 88 pounds (40 kg). Its 7-foot (2.1-m) long ancestor* Diprotodon *may have roamed Australia only 20,000 years ago.*

WOOD ANT

Workers from a* Formica rufa *colony collaborate to move a piece of rotten wood. It will be added to the mound above their subterranean nest.

The wood ants are noted for the huge mounds they build over their subterranean nest, transporting quantities of twigs and pine needles to do so. They live in Europe in conifer forests, in open deciduous woodlands or on tree-studded heathland, according to species. Wood ants, genus *Formica*, are red with a black abdomen and they have a large scale on the waist. The workers are ⅙–⅖ inches (4–10 mm) long, and the queens are ½ inch (12 mm) or longer, the males being only slightly smaller than the queens. The meadow ant or hairy wood ant, *F. pratensis*, regarded by some scientists as a distinct species and by others as a subspecies of the common wood ant, *F. rufa*, is darker in color than the wood ant and more bristly. It nests in more open country. This article focuses on the habits of the common wood ant.

A spreading city

The colonial nest is usually built around a tree stump, but because nests may persist for many years the stump often disappears as a result of decay. The part of the nest above ground is made of small twigs and leaf stalks or pine needles and is really the roof or thatch of the nest, the greater part being underground. The thatch keeps out rain and also keeps the temperature of the nest equable in very hot or cold weather. In an old colony it may be 5 feet (1.4 m) high and 10 feet (3 m) in diameter. Below the mound is a large, rounded pit filled with a mixture of leaf mold and soil. Beneath this, excavated channels slant downward. In winter the ants retire to these channels, hibernating in clusters of several thousand individuals, each cluster having two or three queens at its center. During summer the whole nest is occupied, and open galleries are maintained throughout the mound as well as in the underground part of the nest.

The openings to the exterior are closed with twigs at night and guarded by sentinel ants by day. The worker ants stream in and out with food or materials for the ever-increasing thatch. Visible tracks, from which leaves and other obstacles have been cleared, run out from the nest and can sometimes be followed for over 50 yards (46 m). These lead to foraging sites, trees on which large numbers of aphids are living, and to other nests. The wood ant nests in an area of woodland are usually interconnected, like towns in a county. Some are small and are obviously satellites of a large nest nearby; others have an independent existence, but the ants are not hostile to those from other nests in the vicinity.

Durable dynasties

Inside the nest the workers are concerned with feeding the queens and the larvae, taking over eggs from the queens and transporting larvae and pupae to parts where the temperature and humidity are optimal. The pupae are contained in whitish, ovoid cocoons about as large as the ants. These are the so-called ants' eggs, sometimes used as food for cage birds and fish in aquaria. The real eggs of the ants are microscopic in size. When a pupa is due to hatch, workers tear open the cocoon and help the occupant out. If undisturbed, a wood ants' nest may go on indefinitely: new queens are constantly recruited.

Founding new nests

Wood ant males and queens are winged, but there is no nuptial (mating) flight as in many smaller kinds of ants. Mating takes place on the surface of the nest or on the ground nearby, usually in June or July. After mating, the queen breaks off her wings before doing one of three things. Her usual course is simply to go back into the nest where she lived as a larva and add her eggs to those of her mother, sisters and aunts. Alternatively, she may set out from her home after mating, accompanied by a few workers, to find a site in which to found a new nest. Her third option, if she wanders off alone, is to enter a nest of a related but smaller species, the large black ant *F. fusca*, displace the single queen, and lay her own eggs, which are tended by the *fusca* workers. For a while there are both *fusca* and *rufa* workers in the nest, until the former die out, since there is no longer a *fusca* queen, and thus a thriving new *rufa* nest is founded.

Ant farms

Wood ants are fierce carnivores and bring great numbers of insects, especially caterpillars, back to the nest. Small victims are carried by single ants, whereas large ones are dragged along by teams. The prey is devoured by the workers, which then feed the larvae with liquid nourishment directly from their mouths. Wood ants also tend and milk aphids, genus *Aphis*. on bushes and trees, especially oak and birch. A tree supporting a large number of these aphids is treated as a farm by the ants of a particular nest and is keenly guarded, and a column of ants can be seen running up and down its trunk. To obtain a meal of honeydew, the ant solicits an aphid by stroking it with its antennae. The aphid responds by giving out a drop of its sugary excreta, which the ant licks up. This sugar is the main energy-producing food of the ants, and entomologists estimate that the occupants of a large wood ant nest may collect from aphids the equivalent of 20 pounds (9 kg) of dry sugar in a season.

WOOD ANTS

PHYLUM **Arthropoda**

CLASS **Insecta**

ORDER **Hymenoptera**

FAMILY **Formicidae**

GENUS AND SPECIES ***Formica rufa; F. aquilonia; F. lugubris; F. exsecta; F. pratensis; F. sanuinea***

ALTERNATIVE NAMES
Red wood ant (*F. aquilonia, F. lugubris*); meadow or hairy wood ant (*F. pratensis*); slavemaker ant (*F. sanguinea*)

LENGTH
Up to ½ inch (12 mm)

DISTINCTIVE FEATURES
Red ants with black abdomen and scale on waist; mound-shaped nests of forest detritus

DIET
Various insects; honeydew (aphid excreta)

BREEDING
Males and sexual females mate in summer; males die and females shed wings; each female adds to her natal colony, starts a new one or usurps a neighboring black ants' nest; workers build colony and tend to eggs

LIFE SPAN
Workers: a few weeks; queen: several years; colony: indefinitely

HABITAT
***F. rufa*: open deciduous woodland; others: mainly conifers or heathland**

DISTRIBUTION
Europe

STATUS
Declining

Chemical warfare

Wood ants are well armed for warfare against intruders and other ant species. When threatened, the ant curls its abdomen forward between its legs and squirts a spray of formic acid at its enemy. In fights with other ants it bites with its sharp jaws and then ejects acid into the bite. The strength of the acid can be tested by leaning over a nest and smelling the sharp tang of the acid or feeling it sting the eyes. The green woodpecker, *Picus viridis*, ignores the pain and may sometimes be seen head-down in a nest, licking up the agitated inhabitants with its long, sticky tongue.

At one time, wood ants' nests were regularly raided by animal dealers for the pupae, which were sold to feed pet birds and fish. As long ago as 1880 the German forestry authorities, recognizing the great service that these ants do in destroying insect pests among the trees, passed a law forbidding interference with the nests under penalty of a fine or imprisonment. Today, pet shop dealers are less energetic in collecting ant eggs and there also are fewer woodpeckers, but already there is a noticeable decline in the number of wood ant nests as their wood and heath habitat is usurped for development.

This mound of litter in a Swedish forest marks the site of a wood ant nest. To the residents, it is the equivalent of a city with a population of hundreds of thousands.

WOODCOCK

The plumage of the American woodcock enables it to blend in almost perfectly with its surroundings.

WOODCOCKS BELONG TO the sandpiper family (Scolopacidae) and are most closely related to snipe (discussed elsewhere). They are larger than snipe, up to 13¾ inches (35 cm) long, with brown plumage mottled and barred with black. The straight bill is 3 inches (7.5 cm) in length and the eyes are set well back on the head, giving the woodcock a distinctive expression. The eye sockets are very large, and the openings of the ears lie below the eye sockets rather than behind them. Compared with the snipe, the woodcock has a larger head, a shorter neck and more rounded wings.

Owl-like plumage

There are six species of woodcocks, but only two have extensive ranges. The Eurasian woodcock, *Scolopax rusticola*, breeds in much of Europe, except the extreme north and south, and across Asia to Japan. It is also found in Madeira, the Azores, the Canaries and Asia Minor. Its plumage is very similar in color to that of an owl and makes a perfect camouflage among dead leaves. The back and wings are dark rufous brown, mottled with black and buff, and the crown of the head is dark with light transverse bars. The underparts are buff with dark brown bars. The American woodcock, *Philohela minor*, is confined to the eastern half of North America as far north as the St. Lawrence River and the Great Lakes. It is lighter than the Eurasian woodcock in color and is buff underneath with no barring.

Unlike most waders, woodcocks are solitary birds and it is unusual to see more than two together. They live in coniferous and deciduous woodlands, especially where there are brambles or some other type of undergrowth. They also frequent more open ground, such as heaths with scattered trees or the edges of moors, and the birds have a preference for damp places such as wet hollows in woodlands and marshy areas.

During the day, woodcocks lie up in dry places, among bracken, heather or brambles. They sit very still, relying on their plumage to conceal them and only flying up when almost stepped on. Then they rise swiftly with a great whirring of wings, which has been described as sounding like the ripping of stiff paper. The flight is strong and rapid as the woodcock maneuvers between the trees, but the bird soon settles again. Feeding and courtship flights take place at dawn and dusk. Woodcocks fly just above the treetops or along woodland rides, and with their unmistakable outline it is easy to identify them as they make their regular flights along the same route each evening. In Britain, most woodcocks are sedentary, but in the northern parts of their range they regularly migrate southward in winter, traveling into North Africa and southern Asia. The American woodcock migrates into the southern United States.

Nimble bills

Woodcocks feed largely on earthworms. They also eat ground-living insects and their larvae, including earwigs, caterpillars and beetles. Centipedes and spiders are also taken and occasionally freshwater mollusks. Woodcocks also feed on plant material. The birds usually feed at twilight on damp ground, but in bad weather they may forage in leaf litter and along the shore.

The birds seek earthworms and other animals by probing the soil with the long bill. The tip of the bill is well supplied with nerves, and scientists believe woodcocks locate their food by touch. The bill is thrust into the soil with the mandibles closed, but the tip of each mandible can be twisted outward to seize prey so the whole bill does not have to be forced open against the pressure of the earth. In this way, earthworms can be swallowed without the bill being withdrawn from the soil.

EURASIAN WOODCOCK

CLASS **Aves**

ORDER **Charadriiformes**

FAMILY **Scolopacidae**

GENUS AND SPECIES ***Scolopax rusticola***

WEIGHT
About 10½ oz. (300 g)

LENGTH
Head to tail: 13–13¾ in. (33–35 cm)

DISTINCTIVE FEATURES
Long, straight bill; plumage rufous brown above, buff below, with intricate marbling and barring visible at close range only

DIET
Mostly earthworms and beetle larvae; other insects and some plant material

BREEDING
Age at first breeding: 1 year; breeding season: March–July; number of eggs: 4; incubation period: 21–24 days; fledging period: 15–20 days; breeding interval: 1 year, sometimes 2 broods in a season

LIFE SPAN
Not known

HABITAT
Extensive deciduous or coniferous woodlands with some undergrowth

DISTRIBUTION
Eurasia: U.K. east to Sakhalin and Japan, south to Himalayas

STATUS
Common

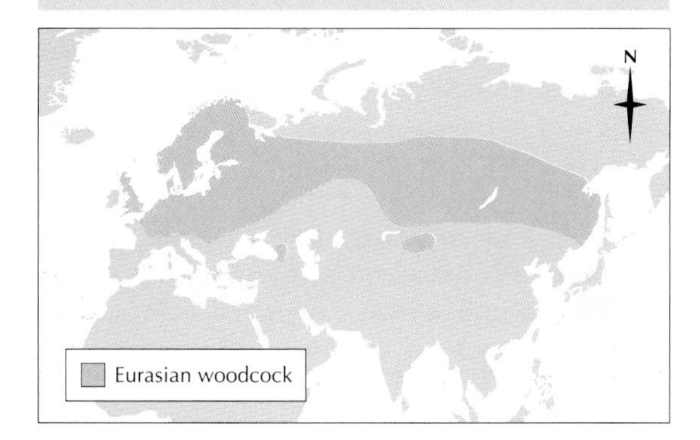

The positioning of the woodcock's eyes gives the bird a greater field of vision to its rear than to its front, so it can watch for danger while it is feeding.

At dusk and dawn during the breeding season, the male woodcock makes special flights around its territory. In Britain, this flight is known as roding. The woodcock flies about 20–30 feet (6–9 m) above the ground with slow, owl-like wingbeats and follows a regular course for up to an hour, starting at sunset. Two sounds can be heard as a roding woodcock flies past. One is a thin, far-carrying whistle, and the other is a low croak as the woodcock checks in its flight. When two roding males meet, they chase each other, and females are also chased. Roding has the function of song in proclaiming ownership of a territory.

The male courts the female by strutting around her with his feathers fluffed out and his wings drooping. After mating, the female makes a depression in the ground, lined with leaves and usually at the foot of a tree. She lays three or four well-camouflaged eggs and incubates them for 21–24 days. Although his plumage is as well camouflaged as that of the female, the male woodcock takes no part in incubation and rearing the young. The chicks fly in about 3 weeks, and there may be two broods in a season.

If a predator approaches, the adult diverts attention from the young by feigning injury. It flies with tail spread and legs lowered or runs about with the tail fanned and the wings drooped or thrashing.

Rides for young

Woodcocks may be among the few birds that carry their young. Jacanas (discussed elsewhere) carry their chicks under their wings, and swans (discussed elsewhere) carry their broods on their backs. Both the American woodcock and the Eurasian woodcock are reported to carry chicks, but reliable evidence is not easy to obtain because of the difficulty of discerning what a woodcock may be carrying as it speeds away. Nevertheless, woodcocks certainly feign carrying their young as part of a distraction display. Perhaps this behavior has fooled human observers as effectively as it confuses predators.

WOODCREEPER

The buff-throated woodcreeper,* Xiphorhynchus guttatus, *is about 8½ inches (21.5 cm) long and occurs from Guatemala south to northern Bolivia and Brazil.

THE WOODCREEPERS ARE small to medium-sized, slender-bodied birds with a dull coloration. They are related to the oven-birds, family Furnariidae, but are placed in their own family, Dendrocolaptidae. Woodcreepers are 5–18 inches (13–46 cm) long and look rather like creepers (discussed elsewhere). The resemblance is due to convergent evolution (the gradual development of certain similarities between two species) brought about by the fact that the two groups of birds have similar habits. Both climb agilely on tree trunks, for example.

There are about 50 species of woodcreepers, many being difficult to distinguish in the field because of the uniformity of their drab plumage and the difficulty of observing them among foliage. The plumage is mainly dull brown, with reddish wings and tail and often light stripes on the head, nape and underparts. There is considerable variation in the size and shape of the bill. It usually is stout, sometimes flattened and chisel-like or down-curved. The most extreme examples occur in the scythebills, genus *Campylorhampus*, also known as the sicklebills, which have a down-curved bill up to 3 inches (7.5 cm) long that makes up about one-third of the total length of the bird. The barred woodcreeper, *Dendrocolaptes certhia*, has an almost straight bill with a hook at the tip, whereas the spotted-crowned woodcreeper, *Lepidocolaptes affinis*, has a slender, pointed bill. All woodcreepers have sharp, curved claws and stiff, woodpeckerlike tail feathers for gripping vertical trunks. They live in tropical America, from Sonora in northwestern Mexico to northern Argentina, and in Trinidad and Tobago.

Solitary with simple songs

Woodcreepers are solitary most of the time, and the habits of most species remain a subject of scientific conjecture. None migrate, with the possible exception of the narrow-billed woodcreeper, *L. angustirostris*, and their weak songs consist of repetitive trills, whistles and harsh notes.

Woodcreepers are woodland and forest birds. The spotted-crowned woodcreeper, for example, lives in forests where the trees are covered in mosses and epiphytic flowers (flowers that grow on the surface of other plants), but it also comes into clearings where there are only scattered trees. Like most other woodcreepers, it can be seen hopping up and around tree trunks as it forages and then flying to another tree and repeating the process.

Woodcreepers do not use the bill as a chisel, as do woodpeckers, family Picidae, but in the same way as creepers, as a probe for searching in crevices in bark or among the plants growing on tree trunks. The birds pull grubs out of their burrows, and sometimes pry off loose bark to reveal other animals. In addition to insects and their larvae and spiders, woodcreepers also take small snails and frogs A few woodcreepers, such as the great rufous woodcreeper, *Xiphocolaptes major*, of Argentina, feed on the ground.

Borrowed nests

Ornithologists assume that woodcreepers nest in cavities in trees, from ground level up to about 30 feet (9 m), and sometimes in earth banks or fallen logs or among the leaves of epiphytes. The birds are unable to excavate their own holes but

TAWNY-WINGED WOODCREEPER

CLASS **Aves**

ORDER **Passeriformes**

FAMILY **Dendrocolaptidae**

GENUS AND SPECIES ***Dendrocincla anabatina***

WEIGHT
About 1 oz. (28 g)

LENGTH
Head to tail: 7–7½ in. (18–19 cm)

DISTINCTIVE FEATURES
Mostly brown coloration; rufous wing patches; buff supercilium (eye stripe) and pale buff throat; straight, grayish bill

DIET
Insects and larvae; other invertebrates taken from trees or flushed by ant swarms

BREEDING
Age at first breeding: 1 year; breeding season: year-round; number of eggs: 2 or 3; incubation period: 17–21 days; fledging period: 18–24 days; breeding interval: 1 year

LIFE SPAN
Not known

HABITAT
Low- and middle-elevation forests to 4,000 ft. (1,200 m)

DISTRIBUTION
Southeast Mexico south through northern Guatemala and Honduras to western Panama

STATUS
Fairly common

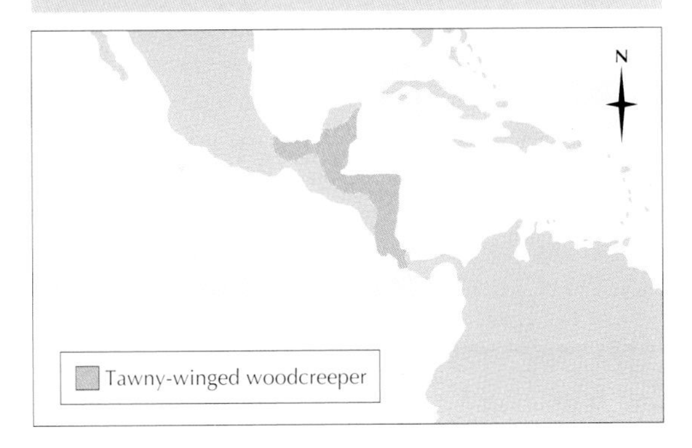

use natural apertures or those of barbets, family Capitonidae, and woodpeckers, particularly if they are well concealed by moss and have narrow entrances. Sometimes, woodcreepers add no new material to form a nest, but at other times pieces of wood and bark from dead trees are deposited in the nest hole, particularly if it is very deep. There is one record of a buff-throated woodcreeper, *Xiphorhynchus guttatus*, carrying some 7,000 pieces of material to its nest. Material is also added throughout the incubation period.

A woodcreeper clutch consists of two, sometimes three, glossy white eggs. The incubation period of the streak-headed woodcreeper, *Lepidocolaptes souleyetii*, for example, is 15 days, and the young stay in the nest for 19 days. Both parents care for the young, incubating the eggs and then brooding and feeding the chicks, which are fed for a few weeks after they leave the nest.

At 6 inches (15 cm), the wedge-billed woodcreeper, Glyphorhynchus spirurus, *is one of the smaller woodcreepers.*

Sharing the food

The range of the woodcreepers is very similar to that of the antbirds, family Formicariidae, and both kinds of birds have the habit of following advancing hordes of army ants (discussed elsewhere), feeding on the small animals that they flush. The woodcreepers fly from perch to perch just above the army ants and swoop out to catch the fleeing insects. Woodcreepers sometimes come into conflict with antbirds, which drive the former from the lower levels of the undergrowth, where it is easier to catch fugitives from the army ants, and also force them to keep to the edges of the ant swarm while they forage in the center. Such competition between unrelated species has an adverse effect on the dominated species only if there is a shortage of food. It appears, however, that the ants flush so many animals that the woodcreepers can find sufficient food even when driven from the best pitches. A large ant swarm may attract a lot of woodcreepers, which depart from their usual solitary or paired behavior. Five or more species may be seen feeding together.

WOOD DUCK

One of the most colorful of North American ducks, the wood duck has been saved from possible extinction due to overhunting and habitat destruction.

THE WOOD OR CAROLINA duck, *Aix sponsa*, sometimes called the tree duck, is probably second only to its close relative the mandarin duck, *A. galericulata*, in terms of ornate plumage. It is 16–20 inches (41–50 cm) long, rather smaller than an eider, *Somateria mollissima*, with a large head, short neck and a long square tail. The bill is short and the toes bear sharp claws.

The breeding plumage of the male is extremely colorful. The feathers of the back have metallic, iridescent colors, mainly green with blue on the trailing edges of the wings. The upper breast is reddish brown and the upperparts are blackish blue. The underparts are buff. The head bears a trailing crest, purple and green with white markings. The eye is red, as is the bill. In its eclipse plumage the male resembles the female but has more white under the chin.

The female wood duck has a small crest on a gray head with a white stripe behind the eye. The neck and breast are brown, speckled with white, and the upperparts are dark brown, with blue on the wings. The underparts are white and the bill is gray.

The wood duck is confined to the United States, the southern parts of Canada, and Cuba. Its close relative, the mandarin duck, lives in eastern Asia, Japan and Britain, and has almost identical habits. Its plumage is even more pronounced than that of the wood duck, having a larger crest and upswept, finlike appendages on the back. The females of the two species are virtually indistinguishable. However, despite their close relationship, the two species do not interbreed. The mandarin does not interbreed with any species, although the wood duck has bred successfully in captivity with several species of ducks belonging to the genus *Anas*.

The Australian maned goose or wood duck, *Chenonetta jubata*, belongs to the same group of perching ducks as the wood duck of North America. It is found over most of Australia and often nests in trees.

Limited migration

The wood duck is sometimes called the summer duck because it nests in the southern states, whereas other ducks merely stay there for the

WOOD DUCKS

CLASS **Aves**

ORDER **Anseriformes**

FAMILY **Anatidae**

GENUS AND SPECIES **Wood duck, *Aix sponsa* (detailed below); mandarin duck, *A. galericulata*; maned goose, *Chenonetta jubata***

ALTERNATIVE NAMES
Carolina wood duck; tree duck

WEIGHT
17½–32 oz. (500–900 g)

LENGTH
Head to tail: 16–20 in. (41–50 cm)

DISTINCTIVE FEATURES
Male: green-and-white head; green crest; vinaceous (wine-colored) breast; black-blue upperparts; buff underparts; red bill. Female: gray head with white stripe behind eye; hint of crest; brown neck and breast, speckled with white; dark brown upperparts; white belly; gray bill.

DIET
Mostly surface aquatic vegetation; insects and other invertebrates in summer

BREEDING
Age at first breeding: 1 year; breeding season: February–May; number of eggs: 9 to 12; incubation period: 27–33 days; fledging period: leave nest after hatching; breeding interval: 1 year

LIFE SPAN
Not known

HABITAT
Lakes, marshes and slow-moving rivers among deciduous or mixed woodland

DISTRIBUTION
Northwestern and eastern U.S.; Cuba

STATUS
Common

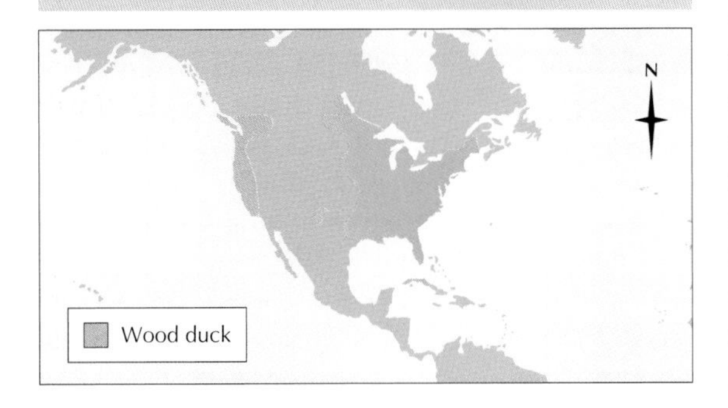

Most adult wood ducks prefer acorns or other nuts to other foods. The picture above shows a male wood duck with its characteristic green head crest.

winter, migrating north in spring for nesting. The wood duck has only a limited migration, a few individuals staying in the northern part of the range for the winter and a few migrating south of the United States. Migrating wood ducks travel in small parties, not mixing with other ducks. Their only calls are whistles. A few vagrants have turned up in Iceland in the fall.

Agile climber

The wood duck is exclusively a freshwater duck, living on ponds, streams and swampy woodland. It roosts on small ponds and flies out in the early morning to its feeding grounds. Unlike many ducks, it is extremely agile in the air and darts through thick forests with remarkable ease, turning and dodging between branches, even at dusk. Wood ducks walk well and climb agilely along branches.

Surface feeder

Wood ducks feed mainly on plants floating on the surface or growing just above the water, although they do upend in shallow water to feed on the bottom. They eat the seeds and leaves of many water plants, including those of grasses, docks, wild celery and wild rice. They also uproot underwater bulbs and tubers. Wood ducks also forage for acorns and chestnuts on the forest floor. In the summer, a considerable proportion of the diet is made up of animal food

The ducklings follow the female wood duck from the nest just 1 day after hatching, often falling some distance without apparently suffering any harm.

taken from around the surface of the water. Two-thirds of this quantity comprises insects, such as beetles, dragonflies and damselflies and their larvae. Snails, crustaceans, small amphibians and small fish such as minnows make up the rest.

Sharing nests

Apart from occasional nests in rock crevices, wood ducks build their nest in holes in trees, either natural cavities formed where a branch has broken off or in the abandoned nests of woodpeckers, flickers and fox squirrels. The ducks are able to squeeze through surprisingly small holes. The hole may be as high as 50 feet (15 m) up or as low as 5 feet (1.5 m), and is usually near water. However if there are no suitable sites, the ducks may be forced to nest 1 mile (1.6 km) from the nearest stretch of water. In these circumstances, there often is competition with the hooded merganser, *Lophodytes cucullatus*, which sometimes results in a duck of each species laying eggs in one nest hole.

No fledging period

Apart from the remains of the previous tenant's nest, the wood duck's nest consists only of a layer of down plucked from the duck's breast. The clutch consists of 9 to 12 eggs, which are incubated for 27–33 days by the female while the male waits nearby. Although there are descriptions of newly hatched wood ducklings being carried down from the nest by the mother, it is more usual for the young to jump down, encouraged by her calls.

There is no fledging period. Shortly after they are hatched, perhaps as little as 1 day, the ducklings scramble up the sides of the nest hole with the aid of their sharp claws and drop to the ground. They sometimes jump as far as 50 feet (15 m) without harming themselves. The parents then escort them to the water, and the ducklings do not return to the nest.

Help for the wood duck

At one time the wood duck was one of the most common and widespread ducks in the United States, but by the early years of the 20th century it was in danger of extinction due to overhunting and habitat destruction. Its numbers had been so reduced that there were fears for its survival in the wild.

Although a fast and agile flier, the wood duck had been shot in vast numbers, while the clearing of trees, especially dead trees, robbed it of nesting places. However, the wood duck, like the mandarin, was highly prized by aviculturalists and the species was saved by careful conservation efforts.

By the time of World War I, there were probably more wood ducks in captivity than in the wild. At this point, naturalist Alain White set up a sanctuary to breed wood ducks. Artificial nest boxes were constructed and placed on top of poles over and around bodies of water. This action was instrumental in halting the decline of breeding populations. By 1939, when the project was ended, more than 9,000 wood ducks had been successfully reared and then released to spread either locally or in other states.

WOOD LOUSE

Wood lice, popularly known in the United States as pill bugs or sow bugs, are particularly interesting in that they are the only crustaceans that have become adapted to a life spent entirely on land. They are small, never longer than ¾ inch (18 mm), with an oval body that is convex above and flat or hollow beneath. The head and abdomen are small, and the comparatively large thorax is composed of seven hard, individual, overlapping plates. There are seven pairs of legs, the last pair appearing only after the first molt. (All insects, by comparison, have three body sections and three pairs of legs.)

Wood lice are scavengers that are found in all temperate and tropical parts of the world. A familiar species in the United States is the pill wood louse, *Armadillidium vulgare*. It has a relatively convex dorsal surface and readily rolls up into a ball when handled, which makes it look like a small, gray pill—indeed, it was once used as a popular medicine. Two wood lice common in Europe are *Oniscus asellus*, which has a brownish body and two rows of yellowish spots on the back, and *Porcellio scaber*, which is bluish gray in color. These two are rather flat and do not roll up into a ball when touched.

Confined to damp, dark places

Unlike other crustaceans, such as crabs and lobsters, the wood louse has wholly escaped its ancestral aquatic lifestyle. This freedom comes, however, at a cost, one that is revealed in another important difference between wood lice and insects. In the outer layers of an insect cuticle is a wax that prevents evaporation of body fluids. Insects also have other ways of conserving water, such as eliminating dry feces. A wood louse has none of these. It readily loses moisture in a dry atmosphere and faces death within about 2 hours unless it can find somewhere damp. It therefore avoids the drying effects of the sun, hiding by day in dark, moist places under bark, stones, piles of leaves or in cracks in the ground, and only coming out into the open at night. Special sense organs in the wood louse's tough outer skin help it to detect moisture.

A wood louse walks more hurriedly over dry terrain, slowing down only when it reaches a damp locale. It has an urge to move away from light, but if caught in a dark place that dries out, it tends to move toward the light. Furthermore, a wood louse tends to aggregate with other wood lice when the air is dry, probably because when bunched together they lose water much less rapidly, thus increasing their odds of survival. Indeed, when one wood louse smells another, it automatically walks toward it for precisely this reason.

There is, however, an exception to this rule of behavior. When a wood louse has been kept for a day in moist surroundings, it is more likely than not to be quite unmoved by the odor of other wood lice. The pill wood louse is hardier than most species and can often be seen walking about in strong sunlight.

The hazards of molting

About 9 or 10 times in its life, a wood louse retires to a quiet, sheltered corner to molt its shell and grow a new one. This is necessary because the shell is rigid and does not grow with the animal. The rear segments of its shell fall off first, and after about three days the front segments follow. This leaves the wood louse fairly defenseless and vulnerable even to members of its own kind, which will eat it without any hesitation.

Like most wood lice,* Armadillidium *is equally at home in a basement or under a garden stone, as long as it remains damp.

How they breathe

If the underside of a wood louse such as *Porcellio scaber* is examined with a lens, a white spot can be seen on the outer plates of each of the first two pairs of abdominal appendages. These white spots are the tufts of fine, branching tubes in the interior of the appendages. The tubes are filled with air and open to the outside by a minute pore. In fact, they represent the beginnings of a tracheal system like the respiratory systems of insects and certain other air-breathing arthropods. It is impossible to suppose, however, that wood lice are in any way closely related to insects, and their tracheae must have evolved independently from those of insects.

Small but independent

The breeding habits of *P. scaber* have been observed, but little is known about the habits of other species. After fertilization by the male, a brood pouch appears as a white triangular patch between the female's forelegs. About 20 or more eggs are laid in the pouch, which becomes slightly distended. The pouch is transparent, so the eggs can be seen clearly through a hand lens. They grow larger and larger until they burst and the embryonic young emerge. About 6 weeks after fertilization, a split appears in the pouch through which the young escape, one at a time, over a period of 2–3 days. Around 1/16 inch (1.5 mm) long, they are pearly white with large black eyespots and six pairs of stumpy legs. They are entirely independent. After all the young have emerged, the female's brood pouch becomes flat again and is sloughed off at the next molt, when a new one grows.

Repelling spiders

Spiders prey on wood lice, notably the species *Dysdera crocata*, identified by its grayish beige abdomen, red cephalothorax and large fangs. A slow-moving spider, *Dysdera* hunts by night; on finding a wood louse, it twists its foreparts to punch one fang into the victim's underbelly and the other through the carapace.

Wood lice defend themselves by secretions from their lobed glands that make them distasteful, although few deter *Dysdera*. The secretions are also used by the wood louse species *Platyarthrus hoffmannseggi* to repel ants, centipedes and other predators.

By huddling together, wood lice reduce the rate at which moisture evaporates through their permeable armor.

WOOD LICE

PHYLUM **Arthropoda**

CLASS **Crustacea**

ORDER **Isopoda**

GENUS **Several, including *Armadillidium*; *Kogmania*; *Ligia*; *Oniscus*; *Platyarthrus*; *Porcellio*; *Schoblia*; *Trichoniscus*; *Tylos***

ALTERNATIVE NAMES
Pill bug; sow bug; slater; roly-poly

LENGTH
Up to ¾ in. (18 mm) depending on species

DISTINCTIVE FEATURES
Small, terrestrial crustacean; oval body; 7 leg pairs; dorsal armor plates

DIET
Various scavenged animal and plant matter

BREEDING
Mated female lays 20 or more eggs and moves them to brood pouch, which later splits to release hatched young; young molt several times; age at first breeding: 15–25 months, depending on species

LIFE SPAN
Up to 41 months in *Armadillium vulgare*

HABITAT
Damp, dark places, e.g., under logs or stones

DISTRIBUTION
Worldwide except for polar regions

STATUS
Common

WOOD MOUSE

Easily distinguished from the house mouse by its large ears, the wood mouse, *Apodemus sylvaticus*, is one of Europe's most common rodents. Although adaptable, it seldom enters houses, unlike its larger cousin the yellow-necked mouse, *A. flavicollis*.

The wood mouse's head and body measure 3–4 inches (7.5–10 cm) long, its tail is 2⅘–3¾ inches (7–9.3 cm) long and it weighs ½–1 ounce (14–28 g). The eyes are large and protruding, and the ears are prominent and translucent. The hind legs are markedly longer than the forelegs. The fur is brown with white underparts and white feet, and there is sometimes yellow on the flanks and a yellow patch on the chest. The yellow-necked mouse is slightly larger at 3¾–4¾ inches (9.3–12 cm) long with a 3–4⅗-inch (7.5–11.5-cm) tail, and the yellow chest patch is extended around the neck. It is noticeably more robust and more agile than the wood mouse.

The range of the wood mouse includes Europe, North Africa and Asia Minor. The yellow-necked mouse has much the same range but is absent from large areas. Even where present, the yellow-necked mouse is less numerous than the wood mouse. The former is more of a woodland animal and extends to higher altitudes than the wood mouse, but broadly the two species are similar in habits and were formerly thought to be two subspecies of a single species. Although they often live as neighbors, there seems to be no hybridizing, and attempts to cross captive specimens have been unsuccessful.

Runways and tunnels

The wood mouse lives in woods, fields, hedgerows and yards, making runways under leaf litter and extensive tunnel systems underground with entrance holes of about 1½ inches (3.8 cm) in diameter. It is more active at night but may be seen by day. It tends to live in clans based on a hierarchy with a dominant male, and when brushwood that has covered the ground for some time is lifted, numbers of wood mice may be exposed. Nests, usually underground, are made of finely shredded grass and are used for sleeping and breeding. Disused birds' nests, even high up in trees, may be used for these purposes, as well as for storing food. Each family has a home range of ¼ acre (0.1 hectare) or more. In field studies wood mice were able to home easily from ¼ mile (0.4 km) away, and in half of test cases from nearly ½ mile (0.8 km) away.

Occasionally small cairns of stones about 2 inches (5 cm) high and the same across are found immediately beside or over the entrance to a burrow without blocking it. The purpose is not known. After a while the wood mice dismantle the cairn and scatter the stones over an area of several square yards. However, if someone deliberately scatters the stones, the wood mice retrieve the identical pebbles and rebuild the cairn over or beside the same entrance.

Wood mice often bound in kangaroo fashion, but unlike kangaroos they bring all four feet to the ground for each leap. They can cover up to 4 feet (1.2 m) with each bound.

Food hoarding

Wood mice eat mainly seeds, grain, acorns, nuts, fruit, buds, seedlings, fungi, insects and snails. They gnaw into hazelnuts, leaving a ragged hole at the blunt end, and gnaw through snail shells. Males eat more insects and less green food than do females, whereas juveniles eat more buds and fungi and fewer insects. A feature of both wood mice and yellow-necked mice is hoarding. This seems to take place whenever a crop of acorns, hazelnuts or holly berries is available. Often a pint to a quart (0.6–1.1 liters) of nuts or berries is found in caches. Such hoards are often incorrectly attributed to squirrels. However, squirrels

In a pinch, the wood mouse can make a meal of almost anything. Where possible, it hoards the surplus in a cache to be eaten later.

The yellow-necked mouse favors mature, deciduous woodland, a habitat it unwittingly improves through its habit of burying tree seeds.

living in the same area as wood mice and yellow-necked mice bury nuts and acorns singly, seldom, if ever, in a hoard. If such a hoard is taken and scattered over an area of ¼ acre (0.1 ha), the mice collect and once again bury all the nuts and acorns within a few hours and in broad daylight.

Successive pregnancies

Breeding begins in March, peaks in July–August, and continues until October or, in mild years, through the winter. Gestation is 25–26 days. Each female may have five litters a year, each of around seven pups (young). The pups are born blind and naked, but their eyes open after 2 weeks. They first leave the nest at 16 days, are weaned at 21 days and are sexually mature at 2–3 months. When they first make excursions from the nest, the young may be seen running behind the mother, holding onto her teats. She may also carry one offspring in the mouth, and does not seem to be slowed down by the close attentions of her offspring.

Multiple predators

The main predators of these two mouse species are owls, especially tawny owls, as well as foxes, weasels, stoats and other small to medium-sized carnivores. Carrion crows take young. Yellow-necked mice defend themselves more fiercely than wood mice and are notably quick to bite when handled by humans.

Winter quarters

Yellow-necked mice come into houses in the fall and leave again at the end of winter. In old houses with hollow walls and plenty of space under the floorboards, they bring in quantities of nuts from distances of up to 300 feet (90 m) and transport them up the walls to the top story of the house. In the quiet of the night a yellow-necked mouse scampering along under the floorboards can sound disturbingly loud.

WOOD MOUSE

CLASS **Mammalia**

ORDER **Rodentia**

FAMILY **Muridae**

GENUS AND SPECIES **Wood mouse, *Apodemus sylvaticus*; yellow-necked mouse, *A. flavicollis***

ALTERNATIVE NAME
***A. sylvaticus*: long-tailed field mouse**

WEIGHT
***A. sylvaticus*: ½–1 oz. (14–28 g).**
***A. flavicollis*: ½–1½ oz. (14–42.5 g).**

LENGTH
***A. sylvaticus*: head and body: 3–4 in. (7.5–10 cm); tail: 2⅘–3¾ in. (7–9.3 cm); *A. flavicollis*: head and body: 3¾–4¾ in. (9.3–12 cm); tail: 3–4⅗ in. (7.5–11.5 cm)**

DISTINCTIVE FEATURES
Bulging eyes, large ears and long tail; brown fur with white underparts; *A. flavicollis* has band of yellow fur across neck

DIET
Seeds, fruits, green plants; invertebrates

BREEDING
Age at first breeding: 2–3 months; breeding season: March–October or longer; gestation period: 25–26 days; number of young: usually 7 (up to 11 in *A. flavicollis*); breeding interval: up to 5 litters a year

LIFE SPAN
2–3 months in the wild

HABITAT
***A. sylvaticus*: woodland and fields.**
***A. flavicollis*: mature deciduous woodland.**

DISTRIBUTION
Europe, North Africa and Asia Minor

STATUS
Not listed as threatened

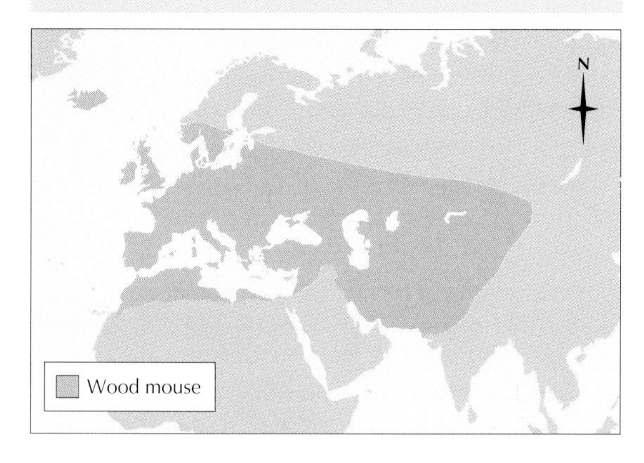

WOODPECKER

No birds are better adapted for a life on the branches and trunks of trees than the woodpeckers, family Picidae. There are some 217 species of true woodpeckers, which occur in wooded areas of Eurasia, Africa and both American continents. Some species have common names other than woodpecker, and two groups of these birds are described elsewhere in this encyclopedia under the separate headings of flicker and sapsucker.

Woodpeckers are up to almost 2 feet (60 cm) in length and usually are brightly colored, with patterns of black, white, green or red. A few woodpeckers have crests. The bill is straight and pointed, the legs are short, in most cases with two toes pointing forward and two facing backward, and the tail is made up of pointed feathers with stiff shafts.

Green and pied woodpeckers

The 15 species of green woodpeckers in the genus *Picus* inhabit the woods and forests of Europe and Asia from Britain to Borneo and Java. The green woodpecker, *Picus viridis*, of Europe is 12 inches (30 cm) long and has a green plumage, which is brighter below, a bright yellowish rump and a red crown. The male has a red-and-black stripe under the eye, whereas the female has a plain black stripe.

The pied, or spotted, woodpeckers, genus *Picoides*, form a widespread group, the 30-odd species being distributed across North America, Europe, Asia and North Africa. They are mostly black or gray with white patches, bars or mottling. The males often have red crowns. The three-toed woodpeckers, which belong to the pied woodpecker group, also have a circumpolar distribution and are unusual in having only three toes on each foot. The ivory-billed woodpecker, *Campephilus principalis*, of North America and Cuba and the imperial woodpecker, *C. imperialis*, of northern Mexico, were two of the world's largest woodpeckers, both dependent on large forest trees. As a result of habitat destruction and disturbance, both species are probably extinct.

A young green woodpecker greets its returning parent. The red-and-black mustachial stripe identifies the adult as a male.

Expert tree climbers

Woodpeckers are usually seen as just a flash of color disappearing among the trees. They live solitarily in woodlands and can be identified by their characteristic undulating flight, three or four rapid wingbeats carrying them up, followed by a downward glide. Instead of being seen, they are more likely to reveal themselves by their harsh or ringing calls, such as the loud laugh of the green woodpecker, or by their drumming, a rapid tattoo that they make with their bills on dead branches or even on metal roofs.

Woodpeckers spend most of their time hopping up tree trunks in spirals, searching for insects. When a woodpecker has examined one tree, it flies to the base of the next and repeats the operation. In climbing vertical trunks, the birds

The acorn woodpecker, Melanerpes formicivorus, *stores acorns in natural or specially excavated holes in trees, the trunks of which can be studded with huge stockpiles of nuts.*

are assisted by their sharp claws, two backward-facing toes and stiff tail feathers, which are used as a prop while climbing.

Boring for insects

Woodpeckers feed largely on insects and their larvae. The green woodpeckers often hunt on the ground for ants and sometimes attack beehives, and the red-headed woodpecker, *Melanerpes erythrocephalus*, of North America catches insects on the wing. Usually, though, woodpeckers feed on insects that are pried out of crevices in the bark of trees or drilled out of the wood. The pointed bill is an excellent chisel, and the skull is toughened to withstand the shock of hammering. When they drill, woodpeckers aim their blows alternately from one side and then the other, in the manner of a tree-feller. The birds remove the insects from the hole by using their extremely long tongue, which, in the case of the green woodpecker, can protrude up to 6 inches (15 cm)

HAIRY WOODPECKER

CLASS **Aves**

ORDER **Piciformes**

FAMILY **Picidae**

GENUS AND SPECIES ***Picoides villosus***

WEIGHT
1½–3 oz. (42–80 g); less in south

LENGTH
Head to tail: 6½–10¼ in. (16.5–26 cm), smaller in south

DISTINCTIVE FEATURES
Longish bill, slightly decurved; mostly black-and-white plumage, black on upperparts and white on underparts; small orange-red mark on crown; middle of back white to brown; outer tail feathers white. Plumage and size vary markedly with subspecies.

DIET
Beetles and beetle grubs, crickets, flies, spiders and vegetable matter

BREEDING
Age at first breeding: 1 year; breeding season: February–June, depending on region; number of eggs: 2 to 5; incubation period: 14 days; fledging period: 28–30 days; breeding interval: 1 year

LIFE SPAN
Not known

HABITAT
Wide range of forests, including Douglas fir (*Pseudotsuga taxifolia*) and juniper (genus *Juniperus*)

DISTRIBUTION
North America, except far north and deserts of southwest; highlands of Mexico and Central America south to west Panama

STATUS
Common

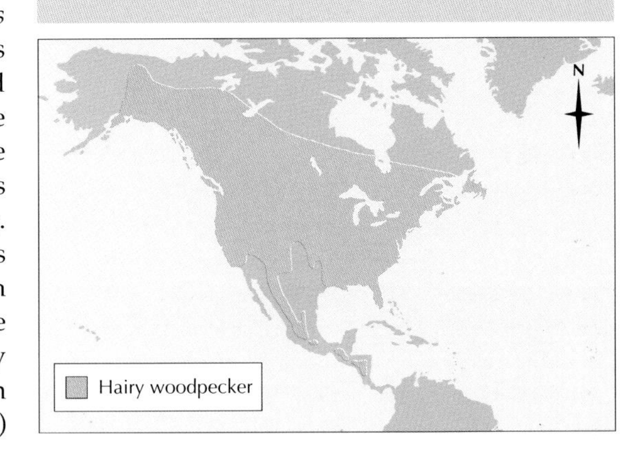

The great spotted woodpecker,* Picoides major*, is a pied woodpecker that occurs in Europe, Asia and North Africa.

from the tip of the bill. The tongue is protruded by the same mechanism as that of the piculet (discussed elsewhere), which belongs to the woodpecker family, and is often tipped with barbs or bristles or coated with mucus for brushing up the insects.

Some woodpeckers eat fruit and seeds or drink sap. The red-headed woodpecker and the acorn woodpecker, *Melanerpes formicivorus*, stockpile acorns, drilling separate holes in trees for each one or using natural cavities.

Nesting in holes

Most woodpeckers nest in holes that they excavate in trees. They drill into a trunk and then tunnel downward to make a cavity up to 12 inches (30 cm) deep. There is no nest lining, and the two to eight white eggs rest on the bottom of the hole. The eggs hatch in 11–17 days and the chicks fledge in 2–3 weeks, depending on the size of the woodpecker. Both sexes bore the nest hole and take turns at incubating and feeding the chicks. Some woodpeckers excavate nest holes in cacti, and some use the nests of social insects. The rufous woodpecker, *Celeus brachyurus*, of Asia, for example, uses the football-sized nests of ants for its main nest site. Other woodpecker species, including the African ground woodpecker, *Geocolaptes olivaceus*, dig burrows in the ground for nesting. The same site may be used over several breeding seasons.

Digging too deep

Boring a nest hole several inches across does considerable damage to a tree and may weaken it sufficiently for it to fall. This was the outcome at the nest of one pileated woodpecker, *Dryocopus pileatus*, in the Everglades National Park, Florida. The tree split off at the level of the entrance to the nest, revealing that the trunk had been hollowed to leave a shell only ¼–½ inch (6–13 mm) thick. F. K. Truslow, a bird-watcher working in the area at the time, concealed himself hoping to observe the reactions of the woodpeckers to this development, which took place during the incubation period. About 10 minutes after the trunk split, the female returned to the tree, disappeared into the nest cavity and reappeared with an egg in her bill. She then flew off with it and did not drop it for the 75 yards (70 m) she remained in view. All three eggs were removed in this manner. Truslow's report of the event is one of the few positive records ornithologists have of birds rescuing their eggs by carrying them away.

WOOD PIGEON

Although the wood pigeon is a heavily built bird, it is a powerful, fast flier, with an estimated cruising speed of 38 mph (61 km/h).

FROM BEING A HARMLESS rarity up to the end of the 18th century, the wood pigeon, or ring dove, has become one of the most common and most destructive pests of agricultural land, especially in parts of Europe. The wood pigeon is a handsome, fairly heavily built bird, up to 16½ inches (42 cm) long with a wingspan of about 18 inches (46 cm). The upperparts are bluish gray, the rump and head being a bluer gray than the rest, while the upper tail is black. The breast is vinous (mauve red), shading to pale gray or lavender on the belly, on the flanks and under the tail. The sides of the neck are metallic purple and green. The base of the bill is pink and expands into a soft, fleshy lump over the nostrils, while the bill itself is yellow shading to pale brown at the tip. The legs and feet are pink with a mauve tinge, and the straw color of the eye and its unusual pear-shaped iris give the bird a very alert expression. The wood pigeon can be distinguished from other doves by the white patch on the sides of the neck, which is absent in young birds, and the broad white band across the wing, which contrasts with the blue-gray inner wing feathers and the dark gray to black outer wing feathers. The male and female are alike except that the males tend to be slightly larger and their plumage is brighter.

The typical race of the wood pigeon, *Columba palumbus palumbus*, is found throughout Europe except in the extreme north. It ranges eastward to Russia and in the south extends to the Black Sea and to the northern coast of the Mediterranean and the various Mediterranean islands, from the Balearics to Cyprus. It is replaced by allied races in northwestern Africa, the Azores and Madeira, and in Asia.

Wary in the country

The wood pigeon is primarily a bird of the woods, but since the spread of agriculture it has taken to feeding on cultivated land. It is also a familiar bird in town parks and suburban gardens, and is often found on downs and on coasts, some distance from woodlands.

From fall to spring, and sometimes also in summer, wood pigeons congregate in large flocks to feed, although single birds and small groups may also be visible. In towns and parks the birds may become fairly tame, but in open country they are wary of humans and take off with a clatter of wings at the slightest disturbance. Their normal flight is fast and strong, with quick, regular wingbeats and occasional glides. On the ground they strut about, restlessly moving their heads to and fro.

Wood pigeons roost in trees, sometimes in large numbers. Their voice, which is heard at all times of the year but more frequently in March and April, is often said to be a series of coos, but the phrase "two coos, Taffy take" repeated several times gives a better idea. The alarm note is a short, sharp *roo* sound.

Agricultural menace

Originally the wood pigeon fed on acorns and beech mast as well as seeds, nuts, berries and the young leaves of many trees. As more and more land has been taken under cultivation and many woods have disappeared, in many areas the bird has turned to cultivated crops to a large extent and found them just as palatable and more abundant. Cereal grains are the most important food

WOOD PIGEON

CLASS **Aves**

ORDER **Columbiformes**

FAMILY **Columbidae**

GENUS AND SPECIES ***Columba palumbus***

ALTERNATIVE NAME
Ring dove

WEIGHT
10–21¾ oz. (284–614 g)

LENGTH
Head to tail: 15¾–16½ in. (40–42 cm)

DISTINCTIVE FEATURES
Plump, medium-sized bird; fairly long tail; blue-gray body; white neck patch; vinous (mauve-red) breast; white wing crescents; broad black tail band

DIET
Mostly plant matter, including leaves, seeds, fruits, buds, flowers and root crops; occasionally invertebrates such as earthworms, snails and insects

BREEDING
Age at first breeding: 1 year; breeding season: March–November; number of eggs: 1 to 3 (2); incubation period: 17 days; fledging period: 33–34 days; breeding interval: 1 year

LIFE SPAN
Oldest ringed bird 16 years 4 months

HABITAT
Primarily woodlands bordered by well-vegetated open spaces, especially in lowlands; parks and gardens

DISTRIBUTION
Most of Europe east to Urals and western Asia, south to Iraq; northwest Africa

STATUS
Common or very common

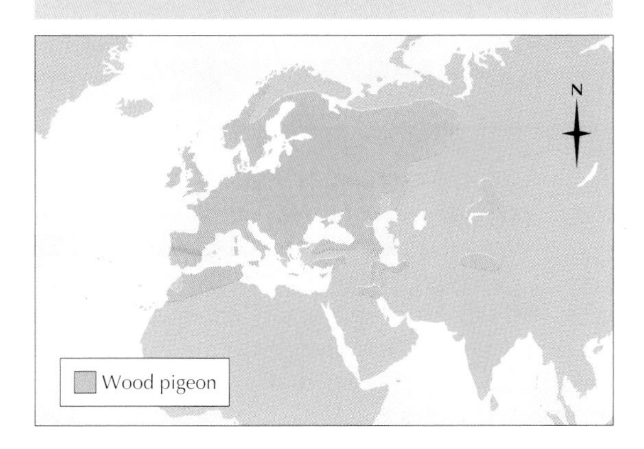

A wood pigeon bathing. Birds bathe as a part of routine feather maintenance and as a way of cooling off in hot weather.

for both adults and young in late summer and fall, and in some areas peas and beans are taken in large quantities. In winter, the birds depend mainly on clover, turnip tops and young greens. The pronounced hook at the end of the bill makes it easy for the wood pigeon to tear off the leaves of these plants. Some animal food is taken, including caterpillars, earthworms, slugs, snails and insects. The wood pigeon needs a fairly large quantity of water and drinks greedily, not in sips like most other birds.

Billing and cooing

The courtship of a pair of wood pigeons begins while they are still living in a flock. A pair separates from the main group and courtship begins. Either on the ground or on a perch in a tree, the birds bow to each other, their breasts touching the ground or perch, with their tails raised and spread, all the time cooing to each other. The bowing and cooing is often interrupted by a nuptial display flight in which the bird rises steeply with strong wingbeats and then glides down and rises again with stiff-set wings. At the top of its rising flight the wood pigeon usually makes several claps with its wings. The sound is caused by a strong downbeat of the wings and not, as so often supposed, by the wings clapping together. Also at this time, pairs of birds start to establish territories in the trees, the males driving away any intruders with aggressive posturing or physical attacks.

Young fed on milk

The breeding season is long, usually from March to November, but in the southern parts of the species' range there are records of birds nesting in every month of the year. The peak of breeding

A wood pigeon broods its young. Uniquely among birds, pigeons produce a form of milk that provides the young with protein that other parent birds deliver in the form of insects.

activity seems to be July, August and September in Britain, a period when there is plenty of ripe corn for feeding the young. There are usually one or two broods a year.

The nest is built in almost any kind of tree or in tall hedgerows, sometimes on top of the old nest of a pair of crows, family Corvidae, or sparrow hawks, *Accipiter nisus,* or on a squirrel's den. Occasionally it is built close to the ground or on a rock ledge. In towns, buildings are used. The nest is a flimsy structure of intertwined sticks, often used for several years in succession. The male brings the material but only the female builds. Usually two, occasionally one or three, white, fairly glossy eggs are laid and are incubated for about 17 days by both parents.

When they hatch, the young are covered in sparse yellow down and for the first 3 days are fed frequently from the parents' crops on a fluid known as pigeon's milk. Ripe cereal grain then becomes the main food, along with some green food and weed seeds supplemented with animal matter. The chicks stay in the nest for about 3 weeks, and afterward are fed by one or both parents for at least another week. The average age attained by a wood pigeon in the wild in Britain is only 38 months but the oldest individual recorded was 16 years and 4 months old.

Large numbers shot

Besides humans, adult wood pigeons have few predators, but many of their eggs are taken by jays, *Garrulus glandarius,* and magpies, *Pica pica.* The losses among young birds are mainly due to starvation, especially when they leave the nest and compete for food with the adult birds. In particularly severe winters, the mortality among wood pigeons is very high but their numbers soon appear to increase again.

Because of the widespread destruction of cultivated crops brought about by wood pigeons, a great deal of research has been undertaken into the optimum methods of keeping down their numbers. Shooting the birds is still the most widely used method, although some sportsmen argue that wood pigeons are difficult to shoot because the shot glances off their feathers. Wide-scale shooting does not appear to make any impression on their numbers.

Migrant or not?

The subject of migration of wood pigeons to and from Britain has provided a constant source of argument among countrymen, sportsmen and bird-watchers for many years. It seems that the wood pigeons in Britain are mainly sedentary but with a tendency to move south in the winter. Only a small proportion of the population undertakes long flights, and these usually are young birds. There probably is a latent urge, inherited from migratory ancestors, that shows itself in only a few individuals. The only birds recovered abroad reached no farther than France. Cold weather movements are sometimes observed, with flocks of birds leaving areas where the ground is frozen or snow-covered to search for better feeding conditions.

In continental Europe, the migratory behavior is rather different. Wood pigeons in Scandinavia and the Baltic are forced to migrate south in winter to escape the snow, and some of these arrive on the eastern coast of Britain, but the numbers vary considerably from year to year. Observers have told of hordes of wood pigeons arriving from continental Europe, and although large numbers may arrive in some years, confusion very often arises because of the flocks of wood pigeons that seem to fly out to sea from Britain and then fly back again. However, irregular arrivals during the fall from continental Europe do take place when birds are drifted over the North Sea as they make their way south buffeted by easterly winds.

WOOD SWALLOW

THE WOOD SWALLOWS, or swallow-shrikes, are an unusual group of birds. They are superficially similar to the swallows but resemble miniature vultures in their habit of soaring. There are about 24 species of wood swallows, comprising a family that is not related to the true swallows. They are 5–8 inches (12.5–20 cm) long, with stout bodies, long pointed wings and short tails. The plumage is soft, and wood swallows are the only songbirds with powder-down feathers, which break up into powder for cleaning the plumage. Other birds with powder-down include the herons, parrots and tinamous. The legs are short and strong and the bill is stout and curved.

The white-breasted wood swallow, *Artamus leucorhynchus*, of Southeast Asia and Australia, is dark brown above with a white bar on the rump. The underparts are white except for a dark brown throat and upper breast. The bill is pale gray. The masked wood swallow, *A. personatus*, is gray above and grayish white underneath. The face and throat are black with a white border. The smallest wood swallow is the little wood swallow, *A. minor*, also of Australia, which is sooty brown with black wings and a black tail with a white tip.

Some species of wood swallows live in Australia. Other species range from the southern Indian subcontinent, southern China and Southeast Asia, where the ashy wood swallow, *A. fuscus*, is found, as far as Fiji, where the white-breasted wood swallow is found.

***Wood swallows are extremely sociable birds and readily huddle together. The picture shows Australian black-faced wood swallows,* A. cinereus.**

Dense huddles

Despite its name, the wood swallow is found not in forests but in open country where it is easily recognizable as it soars in flocks, emitting loud, harsh twittering calls. Apart from ravens and choughs, the wood swallow is the only passerine, or perching bird, that habitually glides or soars for extended periods. It soars in thermals in a manner strongly reminiscent of vultures (discussed elsewhere), sometimes climbing so high that it cannot be seen.

Wood swallows are extremely gregarious birds. The flock usually forages from a particular vantage point such as a tree or, when possible, telephone wires. Unlike true swallows, which perch neatly spaced out, wood swallows huddle together. Some species, for example the dusky wood swallow, *A. cycanopterus*, roost in clumps of up to 200, piled on top of each other. The ashy wood swallow has a distinctive habit of wagging its stumpy tail when it is perched.

Insect eaters

Wood swallows are insectivorous, feeding mainly on flying insects that they catch rather like flycatchers. Individual wood swallows fly out from the perch and snap up insects with their widely opening bills and circle around for some time before returning to the perch. They feed mainly in the early morning and late evening. Wood swallows are specially beneficial to humans because they descend in large flocks, sometimes numbering thousands, onto swarms of locusts and cutworms, the destructive caterpillars of owlet moths. The masked wood swallow is sometimes a nuisance because it feeds on honeybees.

Saucer nests

The nests of wood swallows are usually very loosely constructed saucers of twigs, roots and grass. They are usually placed on a branch,

The white-breasted wood swallow is found in Australia, Papua New Guinea and Malaysia. Wood swallows soar in thermals in the same way that vultures do.

among the bases of palm leaves or in a hole up to 50 feet (15 m) above the ground. The little wood swallow, *A. minor*, nests in colonies among rocks or on cave ledges, and the white-backed wood swallow, *A. monachus*, sometimes uses the solid nests of mudlarks, which it relines. The clutch comprises two to four eggs, and in the species for which information is available, incubation takes about 12 days and is shared by the parents. The chicks fly in another 12 days. There is one record of a brood of young Papuan wood swallows, *A. maximus*, being fed by four or five adults. In such a gregarious bird this behavior may well be relatively widespread.

Food stealer

One report describes two young white-browed wood swallows, *A. superciliosus*, fluttering to the ground to be rescued by the male, which carried them one at a time in his feet to a branch. In *Bird Wonders of Australia* there is an account that shows another side to the wood swallow. An observer recalled watching a newly fledged cuckoo-shrike calling for food when a white-browed wood swallow landed nearby. When the parent cuckoo-shrike appeared with food, the wood swallow grabbed the food from its bill and flew quickly away with the cuckoo-shrike in pursuit. Presumably the wood swallow had somehow learned to associate the begging calls of the young cuckoo-shrike with the arrival of the parent bearing food in the bill.

ASHY WOOD SWALLOW

CLASS **Aves**

ORDER **Passeriformes**

FAMILY **Corvidae**

GENUS AND SPECIES ***Artamus fuscus***

ALTERNATIVE NAME
Ashy swallow-shrike

WEIGHT
About 1½ oz. (45 g)

LENGTH
Head to tail: 6¼–7½ in. (16–19 cm)

DISTINCTIVE FEATURES
Adult: gray head; pinkish buff underparts, brown-gray upperparts; long, pointed wings in flight; black flight feathers and pale underwing coverts; long glides broken by short bouts of rapid wing-flapping. Juvenile: mottled or speckled plumage.

DIET
Mostly insects taken on the wing

BREEDING
Age at first breeding: 1 year; breeding season: January–July; number of eggs: 2 or 3; incubation period: about 12 days; fledging period: about 12 days; breeding interval: 1 year

LIFE SPAN
Not known

HABITAT
Open, wooded country, often with palm trees

DISTRIBUTION
Southern Indian subcontinent, southern China and Southeast Asia

STATUS
Fairly common or common

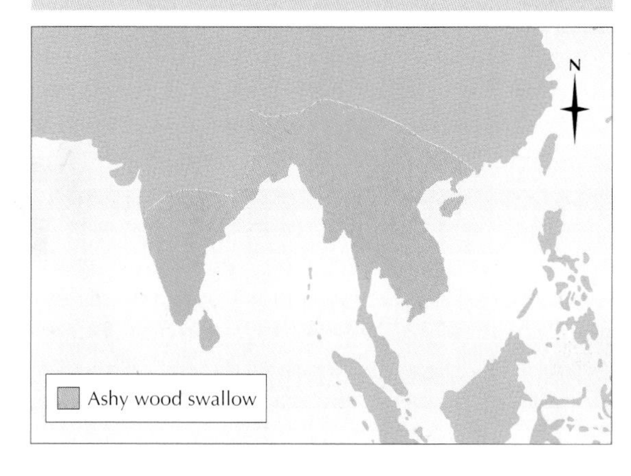

WOOD WARBLER

THE WOOD WARBLERS FORM a large family of small birds confined to the Americas. At present, 116 species are recognized, but it has recently been suggested by some ornithologists that the honeycreepers (discussed elsewhere) belong to the same family and differ only in their adaptations for drinking nectar. Wood warblers range in size from 4–7 inches (10–17.5 cm) long and have narrow, straight bills. The plumage is sometimes dull gray or brown but in many species it is bright, usually yellow, orange or black and white. In tropical America, both male and female are brightly colored, but in temperate latitudes the female's coloration is somber and the male is brightly colored only in the breeding season. The large number of species makes identification difficult; the songs, call notes and male plumage are the most diagnostic features. In Britain, the wood warbler, *Phylloscopus sibilatrix* (discussed elsewhere), is a true warbler in the family Sylviidae.

Wilson's warbler,* Wilsonia pusilla, *from California, United States. Although wood warblers live mainly in woodland and scrub areas, they also inhabit bogs and deserts.

The yellow warbler, *Dendroica petechia*, in which the sexes are very similarly colored, is widespread. It is buff above, yellow underneath, and yellow and black on the wings and tail, the male having rusty streaks on the breast. The yellow-breasted chat, *Icteria virens*, is the largest wood warbler. It is olive green above and bright yellow below, with white around the eyes. Kirtland's warbler, *D. kirtlandii*, with a yellow breast and black-streaked flanks, is confined to an area of 60 by 80 miles (100 by 130 km) in Michigan, where there are dense growths of jack pines, 3–18 feet (1–5.5 m) high.

Two wood warblers have blue in the plumage. The black-throated blue warbler, *D. caerulescens*, is most the most striking of the two, with bluish gray upperparts, black cheeks and throat and white underparts. The golden-winged warbler, *Vermivora chysoptera*, and the blue-winged warbler, *V. pinus*, interbreed where their ranges overlap. The hybrids, which are fertile, were once considered separate species and were called Brewster's and Lawrence's warblers.

Wood warblers breed from Alaska to southern South America, and about half of the species are found in North America and the Caribbean. Like the vireos (discussed elsewhere), wood warblers occasionally get caught up in weather systems that carry them to Europe.

Impressive migrations

Wood warblers are found mainly in woodland and scrub country, but they have colonized a wide variety of habitats. The northern waterthrush lives in bogs, and others are found in deserts and in tropical rain forests.

The chestnut-sided warbler, *D. pensylvanica*, prefers scrub country, and has benefited from the clearing of forests. Most northern wood warblers are migratory, traveling in flocks to Central America and northern South America, sometimes to Brazil and Chile. These flocks, in which several species of wood warblers fly in company with tits, are one of the most dramatic sights of North American bird-watching.

In the spring, migration is rapid, the blackpoll warbler, *D. striata*, taking a month to travel from Florida to Alaska, and flocks of warblers, many in brightly colored breeding plumage, pass through North America. In winter, large flocks of mixed-species wood warblers roam the pine-oak forests of highland regions of Central America and northern South America. The songs of wood warblers are simple when compared with the varied calls of Old World warblers. However, the yellow-breasted chat is a good mimic.

Mainly insect eaters

Almost all wood warblers eat insects, and most feed among the foliage. The waterthrushes, genus *Seiurus*, and the ovenbird, *Seiurus aurocapillus*, feed on the ground, and some wood

A nest of yellow warblers, in Aureolo, Galapagos. Both the male and female yellow warbler are similarly colored.

warblers, like flycatchers, hawk for flying insects. The latter have flattened bills surrounded by bristles for sweeping up their prey.

The black-and-white warbler, *Mniotilta varia*, searches for insects among crevices in bark and has short legs and long claws, which enable it to run up trunks in the manner of a creeper (discussed elsewhere). This warbler is able to stay north in the fall after other wood warblers have migrated because insects hiding in crevices survive longer than those in exposed places. The myrtle warbler, *D. coronata*, survives colder weather because it eats fruit and berries, and can live in areas that have snow in winter.

Varied nest sites

The nests of wood warblers are cup-shaped or domed, some being built 50 feet (15 m) or more up in the tops of trees and others on the ground. The parula warbler, *Parula americana*, builds its nests in hanging skeins of Spanish moss. The nest of the ovenbird, not to be confused with the ovenbirds of the family Furnariidae (discussed elsewhere), is a dome-shaped nest of leaves built on the ground.

The prothonotary warbler, *Protonotaria citrea*, sometimes builds in holes or nest boxes. It is named after the papal secretary, who wears orangish yellow robes. There are usually four to six eggs, the clutch size being lower in tropical America, than in temperate latitudes. The female alone incubates the eggs but both parents feed the chicks. Incubation ranges from 12–14 days and fledging takes about 11 days. The periods are shorter in tropical species than in the northern migratory species.

PROTHONOTARY WOOD WARBLER

CLASS **Aves**

ORDER **Passeriformes**

FAMILY **Parulidae**

GENUS AND SPECIES ***Protonotaria citrea***

WEIGHT
⅗ oz. (17 g)

LENGTH
Head to tail: 5½ in. (14 cm)

DISTINCTIVE FEATURES
Male: brilliant golden-yellow head, throat and breast, contrasting sharply with green upperparts, blue-gray wings, blue-gray tail and white undertail. Female: duller than male, especially on head.

DIET
Insects and spiders; occasionally seeds, fruit and nectar

BREEDING
Age at first breeding: 1 year; breeding season: April–June; number of eggs: 4 to 6; incubation period: 12–14 days; fledging period: 11 days; breeding interval: 1 year

LIFE SPAN
Not known

HABITAT
Flooded or swampy mature woodland for breeding; winters mainly in coastal mangroves

DISTRIBUTION
Breeding: lowland areas of U. S., east of the Rockies. Winter: Central America, Caribbean islands, northern Colombia and Venezuela.

STATUS
Fairly common

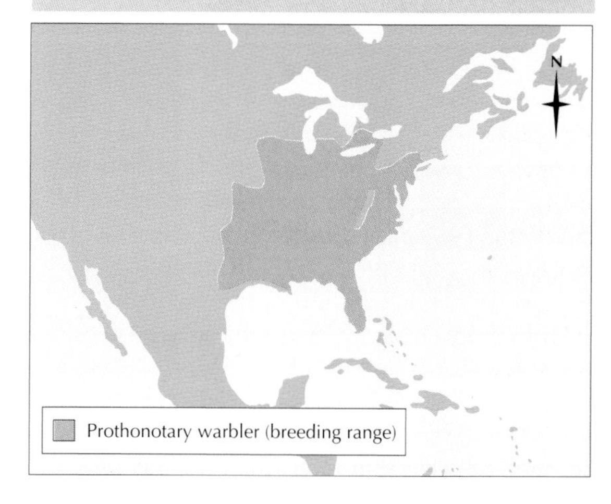

WOOLLY MONKEY

The woolly monkey is commonly kept in captivity, yet very little is known about its life in the wild. It is fairly large and is closely related to the spider monkey (discussed elsewhere). Although only 16–26 inches (40–65 cm) long from head to rump, with a tail 22–30 inches (55–75 cm) long, the woolly monkey is much more thickset than the spider monkey, weighing 8–20 pounds (3.6–9 kg). Like the spider monkey, however, the woolly monkey always seems to be potbellied. It has close, woolly fur and a black face, widely spaced nostrils (a facial feature of all New World monkeys) and a high, rounded forehead. The tail is prehensile (capable of grasping), and the naked undersurface of the tip is covered with ridges and creases like fingerprints. This specialized skin is called volar skin.

Woolly monkeys live in the forests of South America. The more common of the two species is Humboldt's woolly monkey, *Lagothrix lagotricha*, which varies in color in different areas from gray to pale brown or nearly black; often the head is black and the body is pale. Hendee's woolly monkey, *L. flavicauda*, is very rare, and is confined to a small area of montane forests in northern Peru, on the eastern flanks of the Andes. It is a deep mahogany color with a yellow band underneath the terminal half of the tail and a buff nose patch. Hendee's woolly monkey has never been seen alive by Western scientists.

Human activities such as logging and agriculture have adversely affected Humboldt's woolly monkey, which is now an endangered species.

Mealtime acrobats

Humboldt's woolly monkey lives high in the trees, often in the emergent crowns that reach above the forest canopy. It is agile and sure in the branches, partly because of the advantage bestowed by its grasping tail. With the help of this versatile fifth limb, it moves among very thin branches, hanging underneath them and feeding with its hands while grasping only with its tail or with the tail and one foot.

The woolly monkey does not leap, and when it comes to an open space it goes around it or drops into lower branches. It does not seem to come to the ground at all in the wild. In captivity it generally walks in an apelike manner. When resting, it sits upright, grasping a support with its tail. It sleeps at night, curled up, with the tail wrapped around the body or gripping a branch.

Woolly monkeys have been seen in troops of 15 to 25 individuals that appear not to hold territories. As with spider monkeys, membership of a troop is probably temporary. The troops of woolly monkeys often mix while feeding with troops of other monkeys, such as howlers or capuchins. The different species do not compete for food, and they benefit from extra protection against predators, such as birds of prey or large snakes, because there are more eyes to keep a lookout for the silent approach of these killers.

Woolly monkeys feed on leaves, seeds, buds, fruit and occasionally invertebrates. Signs of hard wear on their teeth seem to indicate that they eat a good deal of hard-shelled fruits.

Year-long infancy

There appears to be no fixed breeding season, although there may be a birth peak depending on the locality. The gestation period is 223 days, after which a single young is born. It is nursed for at least a year, reaching puberty at four years. The life span may be as much as 26 years.

Forgotten monkeys

Woolly monkeys are naturally trusting animals that seldom hide from humans. Their habitat is fast being invaded by settlers. As a result, they are highly persecuted in the wild and have been overhunted for their meat and fat. In addition,

Mostly vegetarian, woolly monkeys feed on the leaves, buds and fruits from more than 200 plant species.

these monkeys have suffered at the hands of the pet trade. The mother is generally killed in order to capture her infant. Woolly monkeys cannot adapt to disturbed secondary forest because they are so critically dependent on mature trees. As a result, they are usually the first monkeys to lose out when forest stands are fragmented by logging or agriculture.

Hendee's woolly monkey was first described as *Simia flavicauda* by the German naturalist Alexander von Humboldt (1769–1859) as long ago as 1812, at the same time as he described the more familiar species, which is now named after him. However, Humboldt saw only the trimmed hides of the rarer species, which were being used as saddle covers by muleteers in Peru. He saw nothing of the monkey itself, whose Peruvian name was given as *choro*.

The species was reported again in Maynas District, Peru, in 1832. Then nothing more was heard of it until naturalist Oldfield Thomas of the British Museum (Natural History) in London obtained a woolly monkey skin from the same general area and described it as *Lagothrix hendeei*. He knew of Humboldt's description, but because of its somewhat sketchy nature due to the worn condition of the skins Humboldt saw, Thomas did not consider it to be the same animal.

The matter was dropped again until 1963, when naturalist Jack Fooden pointed out that, in spite of the apparent discrepancies in the descriptions, the two forms must be the same animal, and that moreover it has an exceedingly restricted range in the San Martin and La Lejia areas of the Andes, about 5,000 feet (1,500 m) above sea level. This woolly monkey is now called *L. flavicauda* and is the least known of all South American monkeys.

WOOLLY MONKEYS

CLASS **Mammalia**

ORDER **Primates**

FAMILY **Cebidae**

GENUS AND SPECIES **Humboldt's woolly monkey, *Lagothrix lagotricha*; Hendee's woolly monkey, *L. flavicauda***

ALTERNATIVE NAME
Yellow-tailed woolly monkey (*L. flavicauda*)

WEIGHT
8–20 lb. (3.6–9 kg)

LENGTH
Head and body: up to 26 in. (65 cm); tail: up to 30 in. (75 cm)

DISTINCTIVE FEATURES
Round head; black face; strong, prehensile tail; rotund belly; pale, dark brown or gray to blackish fur varies across individuals and subspecies

DIET
Leaves, buds, fruit; some invertebrates

BREEDING
Age at first breeding: 4 years; breeding season: year-round; gestation period: 223 days; number of young: 1; breeding interval: not known

LIFE SPAN
Up to 26 years in captivity

HABITAT
Undercanopy and emergent trees in mature, primary humid or flooded forests, up to 10,000 ft. (3,050 m) altitude

DISTRIBUTION
***L. lagotricha*: Amazon Basin. *L. flavicauda*: montane forests of northern Peru.**

STATUS
***L. lagotricha*: endangered. *L. flavicauda*: critically endangered.**

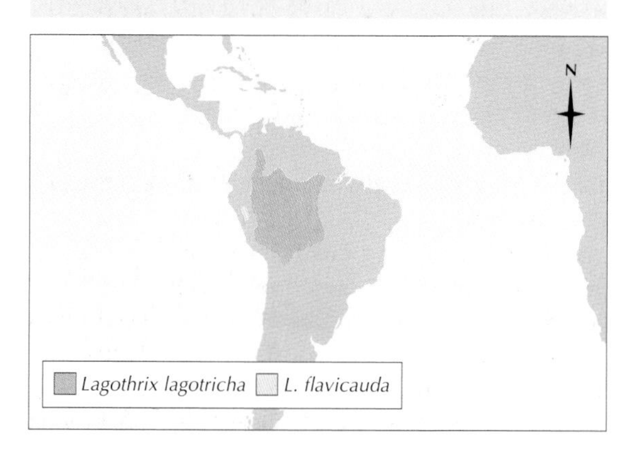

WORM SHELL

Worm shells are an unusual group of mollusks, the shells of which have come to resemble worm tubes. Because they are so remarkable, there is no consensus on what common name to give them. They are variously known as tube mollusks, tube snails and vermetids, as well as worm shells. They start life as fairly ordinary, small, creeping snails with unremarkable coiled shells, but soon they settle and cement themselves to a rock or embed themselves in a sponge. Their shells, which may be various shades of cream, gray, pink or brown, then continue to grow either as long, irregular tubes or in loose coils.

Tube snails are limited to tropical and warm-temperate seas. They are currently divided into two families, the Siliquariidae and the Vermetidae. Scientists suspect that tube snails evolved separately from more typical snails because they differ in a number of important ways.

A worm shell extends its mucus into the sea water around it. Food particles become snagged on the mucus, which is then reeled in and redigested.

Feeding like barnacles

In the family Siliquariidae, the shell becomes a loose spiral, like a corkscrew, embedded in the substrate. The first whorls form a narrow, tight spiral like the turret shells, Turritellidae, from which the group may have evolved. The siliquariids feed by drawing water in over their single gills by the action of gill-mounted cilia (hairlike, mobile organs). Diatoms and other particles become entangled in mucus and are propelled by the cilia to the mouth. In *Siliquaria*, water leaves the body from a slit running the length of the shell. Biologists have observed another method of feeding, unique among mollusks but reminiscent of the feeding of barnacles, in one member of the family, *Stephopoma*. The gill filaments are extended out of the tube and, like the legs of barnacles, are swept through the water to trap small organisms.

Passive breeding

Because worm shells cannot move about, fertilization depends on sperm cells drifting on water currents into the mantle cavity of a female. The eggs of siliquariids are laid in separate capsules and are usually retained, lying free, in the mantle cavity, although in *Pyxipoma* there is a special brood pouch. The veliger (larval) stage is undergone in the egg and the young emerge as crawling snails, not as the free-swimming larvae that might seem more appropriate for the distribution of these otherwise sedentary animals. Like many other snails, the siliquariids have an operculum, a horny lid that can be pulled into place to close the aperture of the shell. It can be unusually elaborate and decorated with branched bristles. When the operculum is half closed, the bristles form a filter, keeping out excessively large particles during feeding. The operculum can be put to another use by the embryo when ready to hatch. Its sharp edge may be used to cut through the wall of the egg capsule.

Boilers in the Bermudas

The members of the Vermetidae have broadly similar breeding habits, except that several eggs are laid in one capsule and the capsules are attached inside the aperture of the mother's shell. The irregular, tubelike shells of some members of this family may be entwined together into masses, sometimes forming large reefs or banks. Together with the tubeworms of the family Serpulidae and the stony corals or madreporians, they form the main constituents of the miniature atolls, or boilers, in the Bermudas and of the reefs off the coast of Israel.

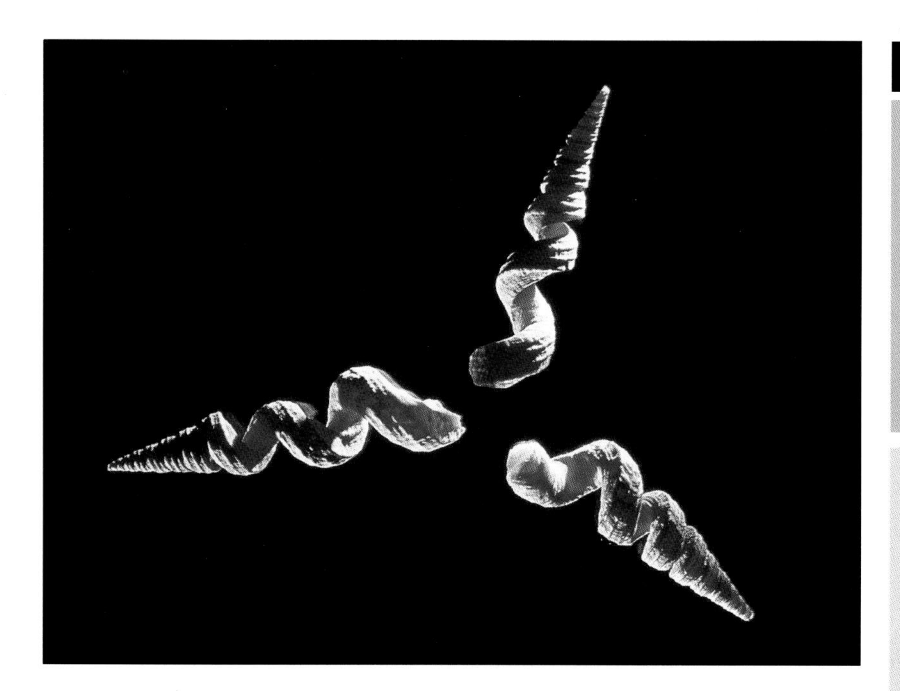

Preserved specimens of Knorr's worm shell,* Vermicularia knorri. *The open whorl is typical and the tightly coiled tip is usually embedded in the substrate.

WORM SHELL

PHYLUM **Mollusca**

CLASS **Gastropoda**

ORDER **Mesogastropoda**

FAMILY **2: Siliquariidae; Vermetidae**

GENUS AND SPECIES **More than 100, including *Vermetus cristatus*, also known as *Dendropoma petraea* (detailed below)**

ALTERNATIVE NAMES
Tube mollusk; tube snail; vermetid

LENGTH
Up to 2–4 in. (5–10 cm)

DISTINCTIVE FEATURES
Shell resembles tube worm; likely to occur in large, reef-forming masses

DIET
Organic particles trapped in water currents

BREEDING
Male's sperm drifts into female on water for internal fertilization; larvae brooded in female's mantle cavity until capable of crawling or swimming

LIFE SPAN
Not known; probably a few years

HABITAT
Shallow waters in rocky or stony areas

DISTRIBUTION
Probably throughout Mediterranean

STATUS
Not listed as threatened

Feeding techniques

The main scientific points of interest in the family Vermetidae center around feeding methods. *Serpulorbis novaehollandiae* lives on the outermost reefs of the Great Barrier Reef off Australia and is one of the few snails to be able to withstand the powerful battering of the waves. Its shell, which can grow to a length of 10½ inches (26 cm), is thick and cemented all along its length to dead coral heads. The mouth, up to 1½ inches (4 cm) across, can be closed by a horny operculum. The species feeds by collecting particles from the incoming stream of water on its long, ciliated gill filaments. The particles are trapped in mucus, which is propelled to the mouth. The mucus is augmented by additional mucus balls from a gland below the mouth.

S. gigas forms twisted shells with thinner walls that grow to 8 inches (20 cm) or so in length, with an internal diameter of up to ⅜ inch (10 mm). Most of the tube is not cemented to the rock, but is raised up and often intertwined with others. *S. gigas* lives only in still water. It feeds in a way that could not work in rough water but which is made easier by the raising of the shell aperture. The gland that in *S. novaehollandiae* merely supplements the mucus passing to the mouth is refined in *S. gigas*, producing threads of slime up to 1 foot (30 cm) long that float in the water. After a while, the threads are devoured as the many little grappling-hook teeth of the radula (a rasplike band) haul them into the mouth, together with any captured food. Larger prey may be taken directly if it passes near the mouth. Strong feeding or respiratory currents would disturb the mucus lines, and the gills are accordingly reduced. Likewise, an operculum would be in the way, and is not present in adults.

Comparing these two species, one may speculate on their evolution from grazing ancestors, the gills of which bore cilia merely to create respiratory currents and sweep away sediment; to creeping snails using their gills for feeding; and then to fixed snails feeding in a similar fashion and producing an excess of mucus that was eventually put to good use as a snare, while the gills reverted to their purely respiratory role. Mucus production has gone a stage further in the scaled worm shell, *S. squamigerus*, of the American Pacific coast. This species secretes not strings, but sheets of mucus up to 6 inches (15 cm) long, and those of neighboring individuals tend to merge into one. Thus, when one worm shell feeds, the others do so too. The record for slime production probably goes to certain Indonesian species that can produce slime strands 6 feet (1.8 m) long.

WRASSE

WRASSES ARE NOTED for brilliant colors and patterns that vary with sex and age. Their belligerence is also exceptional. A typical wrasse is the cuckoo wrasse, *Labrus bimaculatus*, of the western Atlantic seaboard and the Mediterranean. It is long-bodied, compressed laterally, with a long head and jaws. The lips are thick, the teeth in the front of the mouth are well developed and the angle of the jaw extends back nearly to the level of the large eye. The single dorsal fin is long, its front half spiny, the rear half soft-rayed. The anal fin is long, the tail fin is square-ended and the pelvic fins are forward on a level with the base of the pectorals. There are notable color differences between the sexes. The male is yellow or orange with a vivid blue head and back, and with blue lines running over the gill covers and along the flanks. Females and young fish vary in color from orange to red, with three spots on the rear half of the back.

The 500 species of wrasse are most numerous in the Tropics. They range from slender, 3-inch (7.5-cm) fish to 10-foot (3.2-m) giants that weigh several hundred pounds. Some of the small wrasses are cleaner fish that glean parasites or detritus from other fish. One of the most colorful of the larger wrasses is the bluehead, *Thalassoma bifasciatum*, of tropical western Atlantic shores. The male is blue with black bands on the front half and green on the rear, including the crescent-shaped tail. The female and young range from yellow to green. Other Atlantic species are the corkwing or sea pheasant (*Symphodus melops*), Mediterranean rainbow wrasse (*Coris julis*) and ballan wrasse in the east and the pearly razorfish (*Xyrichthys novacula*) toward the Americas.

Home-loving fish

Wrasses are solitary fish that live at most in pairs or trios, seldom in schools. They occur in shallow seas around rocky coasts and coral reefs. Some species live between tidemarks and shelter in rock pools when the tide is out. Another marked feature of their behavior is the tendency for each fish to keep to one area, from which it ventures out to make feeding sorties. This is particularly evident in those species that have become cleaner fish, but with other wrasses, too, observers have noted that they would regularly see the same fish in the same place.

Wrasses swim by twisting movements of the rear part of the dorsal fin and the anal fin, helped by backward beats of the pectorals. When speed is needed, this is supplemented by movements of the whole body, but the fin-swimming allows them to maneuver nimbly in confined spaces, such as the rock crevices they tend to haunt.

Armed with fangs

The strict territorial behavior of wrasses is linked with their habitual aggressiveness. In attacks they typically bite off each other's fins or gouge out eyes. The strong and prominent front teeth, which create havoc in such fights, include the one or more fanglike teeth in the side of each jaw. These are normally used for cracking open crabs and sea snails as well as piercing the flesh of smaller fish. It is thought that wrasses will eat any animal food, including carrion, on the basis that they are often caught in crab and lobster pots, having entered apparently to take the bait.

Rough courtship

Some wrasses construct their nests of seaweed and corallines wedged in a rock crevice, both sexes helping in the work. The eggs are laid in the middle of this tangled mass, but there is no evidence that the parents guard the eggs. In other species, the fish dig a shallow trough in the sand, in which the eggs are laid. Spawning seems to be preceded by a courtship. Observers have noted a male cuckoo wrasse digging a nest in the sand by turning on his side and flapping vigorously with his tail. After that he attacked all the females around him, charging at them and nibbling them until he at last induced one to follow him to the nest. During this excited behavior, the male turned white over a large area of the head and back, which may act as a visual

The humphead wrasse or Napoleon fish, **Cheilinus undulatus,** ***lives in the Indian Ocean and western Pacific. It may grow to nearly 8 feet (2.4 m) and weigh up to 420 pounds (190 kg).***

A corkwing wrasse devours a sea urchin. This fish is found close to shore among rocks or eel-grass beds, and grows to a length of 11 inches (28 cm).

stimulus to the female. The eggs, 1/25 inch (1 mm) in diameter, hatch in 21 days, the fry joining the zooplankton.

Bold colors, bold behavior

The bright colors of wrasses have more than esthetic merit: they raise points about the biology and lifestyle of these fish. First, wrasses comprise one of the few groups of fish in which females and males differ markedly in color. Furthermore, although it is not unusual for fish living around coral reefs to be brightly colored, presumably providing them with camouflage against their background, it is unusual for their relatives in temperate seas to be so brightly colored. Presumably, wrasses have few predators and can be showy with impunity. This idea is borne out by cleaner wrasses, species that remove ectoparasites from often larger, deadlier fish, such as groupers. They could not carry out this trade without an unusual immunity from attack. There are even species of blennies, for example, that mimic the shape and colors of wrasses, a strong indication that these are protective. By contrast, a number of wrasses in the South Pacific look like lumps of floating weed. They are translucent green with dark brown lines and white blotches. Moreover, they hold themselves limp so they are carried back and forth in the surf, as seaweed would be: an effective camouflage.

Sleep tight

Studies of wrasses in marine aquariums reveal the fish spend much of the time resting on rock ledges or in crevices by day, and return to these same places at night to sleep, lying on their flanks to do so. More remarkable are observations of wrasses around Hawaii that bury themselves in the sand at night to sleep. When these wrasses are kept in an aquarium, they swim about all day, and at night the aquarium appears to be empty. Of the 48 species studied, only one does not enter the sand to sleep. This is *Labroides phthirophagus*, which surrounds itself with an envelope of mucus in the manner of a parrotfish. This mucous coating may help prevent the fish's scent from attracting predators.

WRASSE

CLASS **Osteichthyes**

ORDER **Perciformes**

FAMILY **Labridae**

GENUS **60**

SPECIES **500, including bluehead, *Thalassoma bifasciatum* (detailed below)**

LENGTH
Up to 10 in. (25 cm)

DISTINCTIVE FEATURES
Female and young: bright yellow above and white below; prominent black spot on dorsal fin; may flash either broad, green midlateral stripe or a row of squarish green blotches. Territorial male: long, pointed tail lobes; deep green or blue body with black and pale banding. Nonterritorial adult and young males: yellow body color.

DIET
Zooplankton, small animals, ectoparasites on skin of other fish

BREEDING
Territorial adult male spawns in pairs with females of harem; yellow males spawn in groups; eggs pelagic, fry planktonic

LIFE SPAN
Up to 3 years

HABITAT
Reef areas, inshore bays, seagrass beds

DISTRIBUTION
Bermuda south to northern South America

STATUS
Not threatened

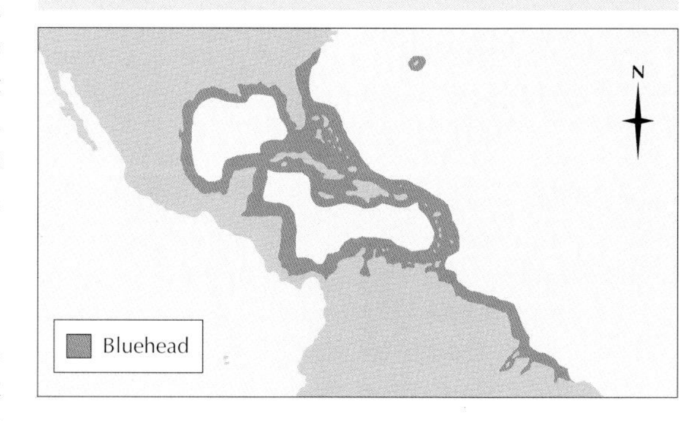

WREN

A winter wren uttering its explosive, trilling song. The generic name* Troglodytes, *meaning cave dweller, refers to this bird's habit of nesting in holes or dark places.

Known in North America as the winter wren, the diminutive *Troglodytes troglodytes* is also a widespread woodland and garden species in Europe. In Britain it is known simply as the wren, occasionally as the Jenny wren, and it is the only European representative of a family, the Troglodytidae, or true wrens, that is otherwise confined to the New World. Birds of other groups also carry the name wren. In Australia and New Zealand there is a large group, which includes the scrubwrens of the family Acanthizidae and the emu-wrens, fairy-wrens and grasswrens of the family Maluridae. A third group of wrens consists of the bush wren, rock wren and rifleman of the family Xenicidae, all three species of which are confined to New Zealand. These Australian and New Zealand wrens fall outside the scope of this article, however, which discusses only the true wrens.

The true wrens are mainly small, drab-colored birds. Exceptions include the cactus wren, *Campylorhynchus brunneicapillus*, of the southwestern United States and Mexico, which is 8 inches (20 cm) long, and the giant wren, *C. chiapensis*, of southeast Mexico, which grows up to 9 inches (23 cm) long. There are about 60 species, the most widespread being the winter wren, which besides being found in North America and Europe also occurs in Asia and North Africa. Up to 4 inches (10 cm) long, in appearance the winter wren has the typical characteristics of other wrens: a squat body, a slender bill, brown plumage that is lighter underneath, and a short, upturned tail and rounded wings, both of which are barred in black.

The house wren, *Troglodytes aedon*, is a very widespread American wren that ranges from southern Canada to Cape Horn at the tip of South America. It is one of three wrens found on the Falkland Islands in the South Atlantic and is very similar in appearance to the winter wren. Bewick's wren, *Thryomanes bewickii*, has a white eye stripe and white underparts, while the cactus wren has a white breast that is densely spotted

Also known as the long-billed marsh wren, the marsh wren occurs from southern Canada south to northwestern Mexico.

with black. The zapata wren, *Ferminia cerverai*, was discovered only in 1926 and is restricted to a single swamp on the coast of Cuba.

The typical habitat of wrens is low, fairly dense undergrowth, but the cactus wren and the rock wren, *Salpinctes obsoletus* (not to be confused with the New Zealand rock wren, *Xenicus gilviventris*), are found in rocky desert country with sparse vegetation. The zapata wren and the short-billed marsh wren, *Cistothorus platensis*, also known as the sedge wren, live in marshes, while several tropical wrens live on the floor of rain forests.

Wrens feed mainly on insects, particularly caterpillars, fly larvae, beetles, bugs and the like, which they find in crevices and on foliage. Cactus wrens turn over stones in their search for insects, and winter wrens have been known to catch small fish.

WINTER WREN

CLASS **Aves**

ORDER **Passeriformes**

FAMILY **Troglodytidae**

GENUS AND SPECIES ***Troglodytes troglodytes***

ALTERNATIVE NAME
Wren; European wren; Jenny wren (archaic)

WEIGHT
¼–½ oz. (7–12 g)

LENGTH
3½–4 in. (9–10 cm)

DISTINCTIVE FEATURES
Adult: tiny, plump bird; brown upperparts; lighter brown, heavily barred underparts; short tail barred with black and held cocked; rounded wings barred with black. Juvenile: similar to adult but less barred.

DIET
Mostly insects, especially beetles; also spiders

BREEDING
Age at first breeding: 1 year; breeding season: March–June; number of eggs: 5 to 8; incubation period: 16 days; fledging period: 14–19 days; breeding interval: 1 year

LIFE SPAN
Usually not more than 2 years

HABITAT
Any that provides low cover, from sea level to 8,000 ft. (2,400 m)

DISTRIBUTION
Most of Europe, parts of northwest Africa, south-Central Asia, North America south to California and Appalachians

STATUS
Common to very common

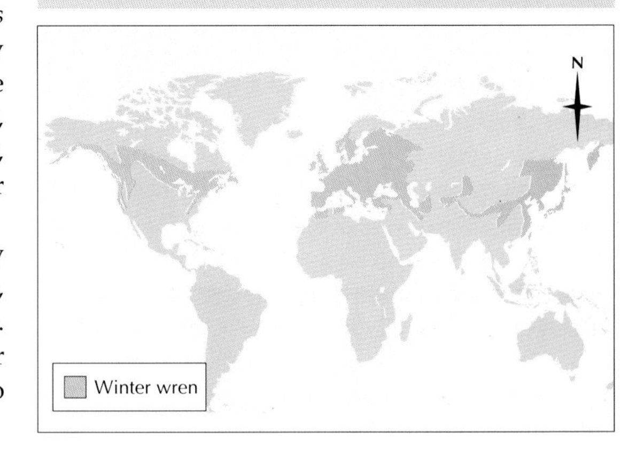

Few wrens migrate, apart from those living in the far north. The winter wren suffers in hard winters when the ground below the undergrowth is covered by snow, and after the severe winter of 1963, when their numbers in Britain were severely reduced, it was possible to determine their preference for different habitats. By means of a painstaking survey over several years, it was shown that the surviving wrens set up territories in woodlands and in well-covered banks of streams. As their numbers recovered, they spread first into orchards and gardens and then into hedges.

Varied songs

Because they usually live among undergrowth, wrens are more often heard than seen. They have a varieties of calls and rich songs, which are surprisingly loud for such small birds. Some tropical wrens sing antiphonally, the pair keeping in contact in dense cover by singing in turn. The song of the winter wren is a shrill warble ending with a trill. It can be heard year-round and is used for various purposes besides advertising the territory. In his monograph *The Wren*, the ornithologist E. A. Armstrong classified the variety of the winter wren's song. The loudest songs are territorial, defying males and attracting females, the harshest song being reserved for when a wren is challenging another bird. Softer, quieter songs are used in courtship and for inviting the female to inspect a nest, and the softest song is heard from females as they sit on the nest. A fairly loud song is also heard when wrens are gathering at a communal roost.

More nests than they need

Most wrens build domed nests, but some, such as the house wren, which often uses birdhouses, build cup-shaped nests in holes and crevices. In monogamous species both sexes build the nest, but in polygamous species, such as the winter wren and the marsh wren, *Cistothorus palustris*, the male builds the nest and the female lines it. These wrens may build several nests in their territories, some of which are ignored or are used only for roosting. These extra nests are sometimes called cock's nests.

In all well-studied wren species, the female alone incubates the eggs, although she may be fed by the male. The clutch size ranges from two to three eggs in the Tropics to 8 to 10 in temperate regions, and incubation takes about 2 weeks. In the monogamous species in particular, the male helps to feed the chicks, which fly after a further 2 weeks. There is more than one brood in a year, and in the case of the house wren the young of one brood may help to feed the young of the next.

Wrens in dormitories

Wrens frequently roost under cover, working their way into crevices, holes in trees, old nests or even pockets of clothes hanging out to dry, and some tropical wrens build special roosting nests. Wrens usually are solitary, but they may gather to roost communally: 61 winter wrens were once counted roosting in a nest box. Communal roosting occurs most frequently in hard weather, and it appears that wrens remember suitable roosting sites for future use. As dusk falls, the wrens gather, and it seems that they follow a leader that calls or sings to them in a particular way. As more wrens fly in to join the group, they chase from perch to perch until they finally enter their roost, arranging themselves tier upon tier with heads directed inward.

The cactus wren,* Campylorhynchus brunneicapillus, *is the largest wren found in the United States. It is the state bird of Arizona.

WRYNECK

THERE ARE TWO SPECIES of wrynecks, the Eurasian wryneck, *Jynx torquilla*, which breeds across much of Europe and Asia, and the red-breasted wryneck, *J. ruficollis*, which breeds in Africa. The wryneck is a relative of the woodpecker but resembles the nightjar in appearance. At a distance the Eurasian wryneck can be mistaken for a thrush, especially when it feeds on the ground. It is about 6¼–6¾ inches (16–17 cm) long, a little larger than a house sparrow, slightly built, with a cryptic plumage of gray crown, mantle and tail, sandy brown throat and breast, and white underparts. All areas of its plumage are delicately marked with black bars and streaks.

Ornithologists believe wrynecks are the most primitive members of the woodpecker family because they lack the stiff tail feathers that woodpeckers use as a prop when climbing, and the bill is weak and incapable of chiseling into trees. However, wrynecks have one toe facing sideways to assist climbing and have long hyoid bones supporting the tongue that run back under the skull and over the top of the cranium, features that are typical of woodpeckers.

Wrynecks do not usually catch insects on the wing. They eat mainly ants, collecting them from tree bark or from cracks in the ground. Pictured is the Eurasian wryneck.

The Eurasian wryneck ranges widely across Europe and Asia from southern England to Japan, except for the extreme north and the extreme southwest and southeast. It winters in parts of North Africa. The plumage is brown, delicately mottled and streaked with gray and black, and the underparts are barred. The red-breasted wryneck of tropical Africa is similar to the Eurasian species, but the throat and breast are rich chestnut and the belly is pale with brown streaks.

Cuckoo's mate

The wryneck's name derives from its habit of twisting its head around over its back when it is alarmed. At one time it was called the snakebird because this action was similar to a snake waving its neck. It was also once known as the cuckoo's mate, because it arrives from its winter quarters in Africa and India at the same time as the cuckoo. Like the cuckoo, it is more often heard than seen, its call of *quee-quee-quee-quee* similar to the shrill notes of a kestrel.

Wrynecks spend the majority of their time in the trees but also feed on the ground, as do some true woodpeckers. Although they lack the spiny tail feathers of a true woodpecker, they sometimes use the tail as a prop. They climb very easily on tree trunks and hide by moving to the other side of the trunk. If wrynecks are disturbed, they fly away fairly slowly on an undulating course.

Dependent on ants

Wrynecks feed on insects, particularly ants and their pupae, and beetles, butterflies and moths and their larvae. Nestling wrynecks are often fed entirely on the pupae and workers of the common black and yellow ants. During dry weather, when the pupae are carried farther below ground, the nestlings may go short of food. Wrynecks occasionally chase flying insects but usually pick up insects from crevices in bark or in the ground or wipe insects off the surface of leaves with their long tongues.

Neck wrigglers

During courtship a pair of wrynecks gape at each other, display their pink mouths and writhe and wriggle their necks in the manner that has earned them their name.

EURASIAN WRYNECK

CLASS **Aves**

ORDER **Piciformes**

FAMILY **Picidae**

GENUS AND SPECIES ***Jynx torquilla***

WEIGHT
1–1½ oz. (30–45 g)

LENGTH
Head to tail: 6¼–6¾ in. (16–17 cm)

DISTINCTIVE FEATURES
Cryptic plumage with black bars and streaks; gray crown, mantle and tail; sandy brown throat and breast; white underparts

DIET
Mostly ants; other insects

BREEDING
Age at first breeding: 1 year; breeding season: April–May; number of eggs: 7 to 10, often 2 or 3 broods; incubation period: 11–14 days; fledging period: 18–22 days; breeding interval: 1 year

LIFE SPAN
Not known

HABITAT
Woodland fringe, open woodland; parks; orchards; large gardens; cemeteries

DISTRIBUTION
Breeding: much of Europe; southern Siberia; north-Central Asia; Sakhalin Island (Japan); southern China. Winter: Africa.

STATUS
Scarce to common

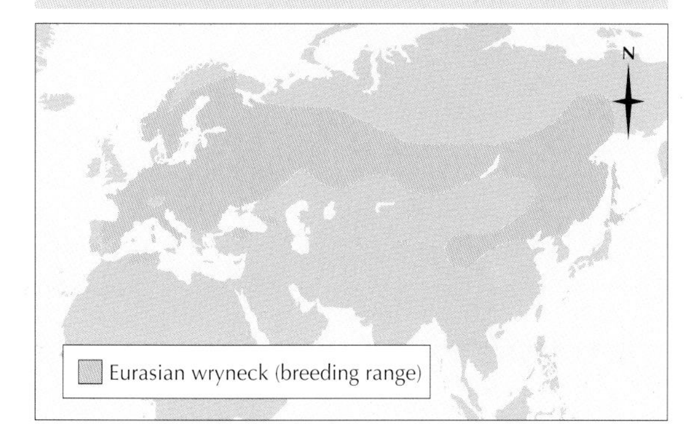

The wryneck has a transparent third eyelid, called a nictitating membrane, which lies underneath the eyelid on the nasal side. It can be drawn across the eye to moisten or clean it.

The usual nest site of wrynecks is a hole in a tree. They are unable to bore their own holes and have to rely on natural ones. They also use bird-houses and holes in banks or in walls. The 5 to 12, usually 7 to 10, dull white eggs are laid on the unlined floor of the hole. Both parents incubate the eggs until they hatch after 11–14 days. The chicks are fed by both parents on insects, nearly always ants, brought in the bill.

Vanishing wrynecks

At one time the wryneck was common enough in Britain to have over 20 local names. However, numbers declined sharply in the second half of the 20th century until in 1981 there was no recorded breeding. The species was therefore effectively extinct in Britain. A decline in wryneck population has also occurred in many European countries, but the British wrynecks have suffered most, perhaps because Britain is at the extreme end of their range.

It is difficult to give a reason for the decline of the wrynecks. Like other hole-nesting birds, they may be short of nesting sites because of land clearance, and increased plowing may have severely affected ant populations. There is, however, no conclusive evidence for this. Whatever the reason, populations of Eurasian wrynecks and some European birds are evidently decreasing, whereas other birds, such as the collared dove (discussed elsewhere), are increasing in numbers. The causes of the changes are often unknown or improperly known, and this makes it very difficult to assess the effects of human activities on animals; a decline in numbers of a species is not necessarily evidence of pollution or overhunting.

X-RAY TETRA

MANY SMALL TROPICAL freshwater fish are semitransparent, but one particular species, *Pristella maxillaris*, has been called the X-ray tetra because it is almost totally transparent. The swim bladder and much of the skeleton can be seen clearly, but the stomach and intestine are opaque and the hind part of the body is translucent.

X-ray tetras are 1¾ inches (4.5 cm) long. The body is fairly deep and ends in a forked tail fin. The first dorsal fin and the long-based anal fin are high. The second dorsal fin is very small, and the pectoral and pelvic fins are of only moderate size. Although the body is usually described as transparent, it sometimes appears silvery in reflected light. At other times the scales have a faint yellowish or greenish tinge and there is a black shoulder spot. The first dorsal fin and the anal fin are lemon yellow with a prominent black blotch or band and they are white at the tips. The tail fin is reddish.

The X-ray tetra lives in the rivers of northeastern South America, from the Amazon Basin to the Orinoco River and the Guianas.

Rapid swimmers

These small fish live mostly in shoals that swim rapidly backward and forward. When in smaller groups, they are rather shy and tend to hide among water plants or in other shaded places. They swim in a jerky manner, letting the tail drop slowly and then flicking the fins to bring it up and to let it drop again.

Early in the 20th century the X-ray tetra became a favorite aquarium fish, popular not only for its semitransparency—which was something of a curiosity—but also for its colors and rapid movements.

Sawlike teeth

The generic name of this fish, *Pristella*, means little saw. Sawlike teeth and an upward-sloping mouth are a clear indication of a carnivorous diet. The X-ray tetra is closely related to the piranha (discussed elsewhere), which has an exaggerated reputation for savagery. The X-ray tetra, being so much smaller, feeds on smaller prey, mainly animal plankton and insect larvae, but it will take anything living of appropriate size, including small worms. Even when young it feeds on animals such as rotifers, the nauplius larvae of crustaceans, small insect larvae, very small worms and the hosts of microscopic protozoans, which used to be called infusorians, that live in the freshwater plankton.

Distinctive sexes

The female is the more robust of the two sexes, the male being noticeably slimmer. The sexes can be told apart by viewing them against the light:

The X-ray tetra is popular among aquarists because of its robustness, its peaceful character and its delicate colors.

X-RAY TETRA

CLASS **Actinopterygii**

ORDER **Characiformes**

FAMILY **Characidae**

GENUS AND SPECIES ***Pristella maxillaris***

ALTERNATIVE NAMES
Water goldfinch, albino pristella, pristella tetra

LENGTH
1¾ in. (4.5 cm)

DISTINCTIVE FEATURES
Near-transparent, greenish yellow body that has a silvery look in some lights; black shoulder blotch; lemon-yellow dorsal and anal fins, each with large, black blotch and white tips; reddish forked tail; female more robust than male, with rounded body cavity when viewed against light; male slimmer, with tapering body cavity

DIET
Worms, small crustaceans, insects

BREEDING
Number of eggs: 300 to 400; hatching period: 20–28 hours; larvae free-swimming after a couple of days; young attain adult colors in around 6 weeks

LIFE SPAN
Not known

HABITAT
Calm coastal waters and densely vegetated swamps

DISTRIBUTION
Amazon, Venezuela and the Guianas in South America

STATUS
Not listed as threatened

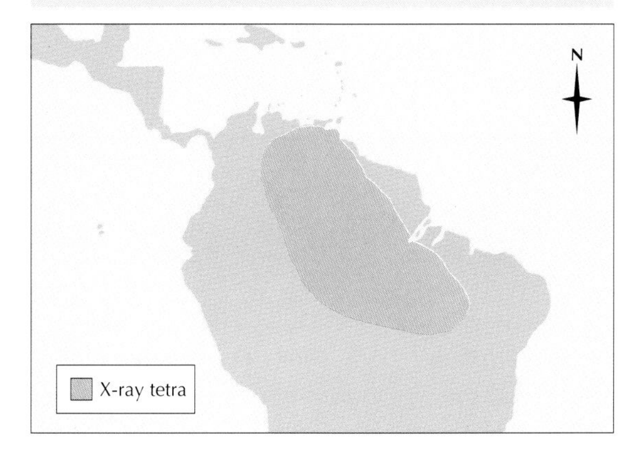

One of the common names for the X-ray tetra is the water goldfinch, inspired by the combination of colors in the fins.

the hind end of the body cavity of the female is rounded, whereas that of the male tapers to a point. In spawning, the eggs and milt (sperm-containing fluid) are shed into the water, where fertilization occurs. Each female lays 300 to 400 eggs at a time. Spawning usually takes place in full sunlight in the morning among water plants. The eggs hatch in 20–28 hours, the larvae hanging like minuscule glass rods on the leaves of the water plants for a day or two before they begin to swim. They keep to the cover of vegetation, however, for the first 2 weeks. The larvae grow rapidly and develop their full colors in about 6 weeks.

Cannibalism

There is little precise information on the natural predators of the X-ray tetra, but one can be fairly sure from the relatively small number of eggs laid that the number of predators is also small.

From what is known by keeping these fish in aquaria, the likelihood is that the main dangers are in early infancy, when the larvae may be eaten by adults of their own kind. Care must be taken to remove the parent fish to another tank after they have spawned.

Popular curiosities

It is always interesting to speculate about the reasons why aquarium fish are popular. X-ray tetras attract attention because they are colorful and lively. However, when they were first displayed in the aquarium of the London Zoo during the early years of the 20th century, the visitors' comments suggested it was first and foremost the transparency of the fish that captured attention. This may have been because people could see inside a fish for the first time and view its internal organs, something that is impossible with many more familiar fish.

YAK

The yak is able to survive in the harsh conditions of the high mountain ranges of the Tibetan plateau and the Himalayas. The yak above is standing next to a Buddhist shrine.

THE YAK IS THE WILD AND also the domestic ox of the Tibetan plateau. It lives in what may be the most inhospitable habitat of any large, hoofed mammal, among vegetation so scanty that its mere survival seems miraculous.

The wild yak, *Bos mutus,* may be the largest of all cattle; bulls may be 7 feet (2 m) high and weigh up to 2,200 pounds (1,000 kg). Domestic yak, *B. mutus grunniens,* are far smaller. In both forms of yak, the shoulders are humped, and a long, shaggy fringe of hair lies on the shoulders, flanks, thighs and tail. The coat is black in the wild yak. In the domestic form it varies from black to piebald or white, and the hair is usually longer. The long, thin and angular horns point straight out sideways, then turn up, and at the tips turn back and usually inward, though in some domestic yak they turn outward. The horns may grow to over 3 feet (1 m) in length.

Domestic yak are found all over the highlands of Central Asia, on the Tibetan plateau and the Himalayas, from the Rupshu plateau in Ladakh across Tibet to Szechwan, in the Pamir, Tienshan and Altai ranges. Sometimes they are crossed with domestic oxen to produce a creature called the dzo, which is bred mainly in Ladakh, at lower altitudes, and is difficult to distinguish from the true yak. Polled yak have been bred in places, and dzo, too, may be hornless.

Wild yak are confined to the highest altitude zones in Tibet and Szechwan, going up to 20,000 feet (6,000 m), in one of the loftiest habitats inhabited by any animal in the world.

Although they are legally protected in China and India, yak have undergone a dramatic decline in the past few decades, due mainly to uncontrolled poaching by local tribes. Wild yak numbers have also been reduced by habitat alteration, disease and hybridization with domestic cattle. Wild yak may have also been displaced from some areas by domestic yak, which compete for the same food. However, herders value wild yak for improving the domestic stock from time to time.

YAK

CLASS **Mammalia**

ORDER **Artiodactyla**

FAMILY **Bovidae**

GENUS AND SPECIES **Wild yak, *Bos mutus*; domestic yak, *B. mutus grunniens***

ALTERNATIVE NAME
Dzo (domestic yak)

WEIGHT
Male up to 2,200 lb. (1,000 kg); female up to 750 lb. (340 kg); domesticated forms smaller than pure forms

LENGTH
Head and body: up to 10 ft. (3 m); shoulder height: up to 7 ft. (2 m)

DISTINCTIVE FEATURES
Massive head and body; humped shoulders; long, thin angular horns, sometimes more than 3 ft. (1 m) long; long, shaggy coat; black coat in wild forms, black, piebald or white in domesticated forms

DIET
Grasses, herbs and lichens

BREEDING
Age at first breeding: not known; breeding season: April–May; number of young: 1; gestation period: 258 days; breeding interval: 2 years

LIFE SPAN
Up to 23 years

HABITAT
Around snow line in steppe and mountainous areas (alpine tundra and ice desert)

DISTRIBUTION
Tibetan plateau and Himalayas

STATUS
Vulnerable

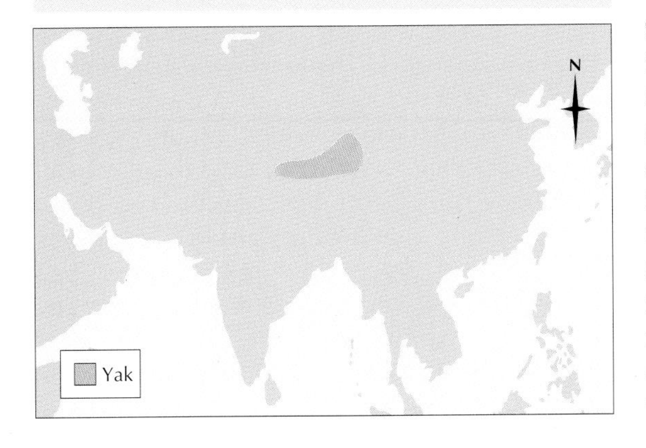

Domestic yaks are smaller than the wild forms. Pictured is a domestic male yak grazing in the high pasture of Sagamartha National Park, Nepal.

Mountain movements

The cows and calves of wild yak live for most of the year in herds of 20 to 200 individuals. The adult bulls are solitary or live in bachelor herds of up to 10 individuals. The herds increase in size in the spring, when the grass starts to sprout. In the summer, the wild yak move higher up the mountains, returning to the lower plateaus in the fall, where they wade through waterlogged valleys in search of moorland grasses.

When the snow comes, around September in Szechwan, the yak make their way once more to the high ground, and there they wander along the snowy upland moors in temperatures as low as -40° F (-40 °C). The herd moves through the snow in single file, each animal placing its feet in the tracks of the one in front. When disturbed, the herd flees at a gallop, the yaks' long flank fringes swaying from side to side. Though immensely heavy, they are surprisingly sure-footed on the rocky slopes. Their only predators are wolves and bears, which may overpower unprotected calves.

Wild yak bulls live alone or in small herds of up to 10 individuals. Rutting (mating) occurs around July, when the herds separate and the bulls compete for cows.

Adaptable animals

Although the yak is so well adapted to cold climates, it has been kept successfully in the subtropical environment of the Rio de Janeiro Zoo, Brazil, with its summer temperatures of 104° F (40° C). In this climate, yak grow shorter coats and their respiratory rate increases.

Fights for females

Yak herds begin to split up in the rut (mating season), which takes place in July in Szechwan but earlier in Ladakh. The big bulls fight for possession of cows, pushing against each other with their foreheads but doing no real damage. Gestation lasts for around 258 days, and the female gives birth to a single young in April or May. The calf is weaned after a year or so, and the female breeds every other year.

Economical yak

The economy of the peoples of the Central Asian mountains is still largely agricultural, and it depends heavily on the domesticated yak. The dzo is used as a beast of burden by growers of barley in Tibet. Its rich milk is made into curd and ghee, an oily butter used widely in cooking. The ghee is mixed with ground barley, or *tsampa,* to produce a staple Tibetan foodstuff. The meat of the yak is eaten in every form, from boiled to dried or fried. The skins and fibers are used as cloaks and for tents, effective in the subzero temperatures on the high plateau. They are also twisted into strong rope. The tail is traditionally used as a fly whisk or broom. The main use of the yak is, however, as a pack animal. Pure yak are used in the highlands, but at lower altitudes dzo are used to carry trade goods, between Ladakh and Kashmir, for example. A mature dzo can carry a load of 300 pounds (136 kg), even over the roughest ground. The strict lamaistic form of Buddhism practiced in Tibet forbids the killing of yak under normal circumstances, although sick yak may be slaughtered humanely.

In the competition for food between wild and domestic yak, the latter, with human interests on their side, always win. As a result, wild yak have been driven into the highest and most inaccessible places in the Himalayas and the Tibetan plateau. Their original range was far wider. In the 14th century, wild yak occurred in the Tuva chain in the Russian Federation. In the 17th century, they were known in Kusnetsk; in 1739 there were wild yak in the Altai and Dauriya, on the western and eastern borders of Mongolia respectively, and in the Semipalantinsk area of Kazakhstan. The northern yak, probably ancestral to the domestic yak still used in the southern Altai, have been described as a separate species, *B. baicalensis,* now believed to be extinct.

At the turn of the 21st century, the World Conservation Monitoring Center, Cambridge, England, concluded that yak numbers had declined severely over the previous two decades, mostly due to illegal hunting. The total population, estimated at 15,000 in 1995, was thought in 2001 to include only 10,000 mature animals.

YELLOW-EYED PENGUIN

In many respects, the yellow-eyed penguin provides a contrast to the other penguins in this encyclopedia. Of the 17 penguin species only six breed in the Antarctic and only the Adélie penguin, *Pygoscelis adeliae,* and the emperor penguin, *Aptenodytes forsteri,* (both discussed elsewhere) are completely confined to the Antarctic, whereas the king penguin, *A. patagonicus* (discussed elsewhere), reaches only the fringes. Although the Antarctic is usually thought of as the home of penguins, the remaining 11 species breed around the shores of Africa, Australia and South America, reaching as far north as the Galapagos Islands on the equator. The yellow-eyed penguin breeds on the coast of South Island, New Zealand, and on Stewart, Auckland and Campbell Islands. In marked contrast to the Antarctic penguins, its nests are well scattered among thick vegetation.

The yellow-eyed penguin is medium-sized, about 26 inches (66 cm) long and, like many penguins, has yellow plumes on the head. The plumes of the yellow-eyed penguin are only moderately elongated, unlike the flowing head-dresses of the macaroni penguin, *Eudyptes chrysolphus,* and the royal penguin, *E. chrysolphus schlegeli.* The crown and nape are gray brown with black flecks, and there is a yellow band running from the base of the bill, through the eye to the nape. The upperparts are gray brown with off-white underparts and white edges to the flippers.

Long-term studies

The yellow-eyed penguin is distinguished by being the first seabird to have its breeding habits intensively studied. Long-term studies have now been made on many birds, but few cover as long a period as that of the yellow-eyed penguin, which was started in 1936 and finished in 1953. Furthermore, it was carried out by amateur ornithologist L. E. Richdale, who worked alone in his spare time. His study was made on a small colony, averaging about 40 nests, in which all the penguins were marked with individually numbered metal rings so their breeding success, fidelity and other aspects of their social life could be studied.

Unlike the Antarctic penguins, but like many of those penguins living in warmer waters, yellow-eyed penguins do not migrate. Instead, they stay in shallow coastal waters near the breeding places year-round, coming on shore to roost. They also molt onshore just after the breeding season, the nonbreeding penguins doing so first.

The breeding penguins have to build up reserves of fat after the breeding period to survive the 3-week molting period, during which they lose 40 percent of their body weight. Like other penguins, yellow-eyed penguins feed on small fish and squid.

Nests in holes

Not all penguins nest in dense colonies on bare rocky cliffs and shores. Some nest in burrows or crevices. The yellow-eyed penguin nests in holes, among rocks or under fallen logs in thick scrub or forests, up to ½ mile (800 m) from the sea. The nests are well spaced out and one nest may be 300 feet (90 m) or more from its neighbors. This is in marked contrast to the colonies of the Adélie penguin, where scientists have found that more chicks are reared in the middle of large colonies, where the nests are packed closely together. The advantage of having nests more spaced out is that there is less fighting between neighbors.

The world population of yellow-eyed penguins numbered only 5,000 birds in 1993.

Most pairs of yellow-eyed penguins stay together and breed in the same nest site year after year.

However there is still some fighting in a colony of yellow-eyed penguins, particularly among young birds establishing nest sites.

The clutch of two white eggs is laid on a nest of sticks and grasses. Most clutches are started in the third week of September and are incubated by both parents for 42–50 days. The chicks are then fed at the nest for another 100–110 days, until they have fledged. This is a long period compared with the 63-day fledging period of the similar-sized Adélie penguin. When they leave their nests, the young penguins, which can be distinguished only by the yellow bands on the sides of the head, gather on pathways leading down to the sea and take to the water together.

Some yellow-eyed penguins breed when they are 2 years old, whereas others do not breed until 3 years. In Richdale's colony, five of the 36 penguins lived for more than 18 years.

Faithful partners

Richdale's study of the yellow-eyed penguin was one of the first to show that some birds return to the same nest site and partner year after year. This is particularly marked in seabirds that are generally long-lived.

The rate of separation each year among the yellow-eyed penguins was about 14 percent, excluding pairs that dissolved through the death of one member, and one pair remained faithful for 13 seasons. A similar separation rate exists in Adélie penguins. Moreover, scientists have discovered that two birds are less likely to remate if they were unsuccessful in breeding in the previous season. In other birds, the separation rate may be lower: for example, in skuas the rate is as low as 2 to 3 percent. In the Adélie penguin, the degree of faithfulness to a mate exceeds that of two birds returning to their old nest site. Individuals can definitely recognize each other by voice, if not by sight.

YELLOW-EYED PENGUIN

CLASS **Aves**

ORDER **Sphenisciformes**

FAMILY **Spheniscidae**

GENUS AND SPECIES ***Megadyptes antipodes***

WEIGHT
11½ lb (5.2 kg)

LENGTH
Head to tail: 26 in. (66 cm)

DISTINCTIVE FEATURES
Medium-sized bird; gray-brown upperparts; off-white underparts; yellow band from base of bill through eye to nape; gray-brown crown; bluish red bill; white edges to flippers

DIET
Small fish and cephalopods

BREEDING
Age at first breeding: 2–3 years; breeding season: September–October; number of eggs: 2; incubation period: 42–50 days; fledging period: 100–110 days; breeding interval: 1 year

LIFE SPAN
Up to 20 years or more

HABITAT
Nests in temperate forests and on grassy coastal cliffs; feeds offshore

DISTRIBUTION
South Island, New Zealand, and Campbell, Stewart and Aukland Islands

STATUS
Localized and scarce

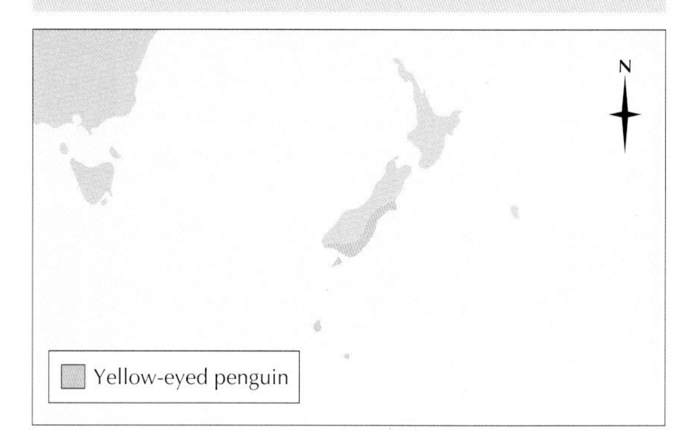

YUCCA MOTH

Two genera of small North American moths pollinate the flowers of yucca plants, and their larvae feed on the plants' seeds. The mutually dependent relationship between moth and plant is unusually intricate and precise. It is one of the best-known cases of balanced symbiosis, an association of two wholly distinct organisms for their mutual advantage. The larvae of the moth can feed only on developing seeds of yucca, and the plant is pollinated only by this particular moth. The moth provides for the needs of both her offspring and the plant by what appears to be a deliberate course of action on her part.

The yucca plants themselves grow in Mexico and the United States and are sometimes called Spanish bayonets, the common name for *Yucca baccata*, inspired by the cluster of dark green, swordlike leaves that spring from the rootstock. A tall stem bearing the flowers rises from out of the center of the clump of leaves. The yucca is a popular plant in ornamental gardens in Europe as well as North America.

The moths belong to the genera *Tegeticula* (formerly known as *Pronuba*) and *Parategeticula*, and are called yucca moths. They are small white moths, usually about ⅖ inch (1 cm) long and about 1 inch (2.5 cm) across the spread wings. The best-known species of yucca moth is *Tegeticula yuccasella*. Some of the different species of moths feed on one particular type of yucca plant. For example *T. yuccasella* pollinates and feeds on *Yucca filamentos*, whereas the Joshua tree, *Y. brevifolia*, and the chaparral yucca, *Y. whipplei*, are associated with *T. synthetica* and *T. maculata*, respectively. This type of association, known as monophagy, was first discovered and described by the American entomologist C. V. Riley in 1872. Several other species of yucca moths feed on a relatively small number of closely related yucca moths, in a process known as oligophagy.

Precision pollination

The yucca moth becomes active after dark, and when she has mated, the female seeks out yucca flowers to collect pollen from them. To do this she uses a pair of curved tentacle-like structures, which are modified maxillary palps. The adult moth never feeds and does not have a proboscis (extended tubular appendage near the mouth) of the normal type. The pollen is sticky and the moth gathers it from open anthers and works it into a ball that may grow to be larger than her own head. Holding the ball with her tentacles, head and forelegs, she flies to another flower of

***The yucca moth and yucca plant are mutually dependent. Should one species become extinct, the other would also die out. Pictured is* T. yuccasella.**

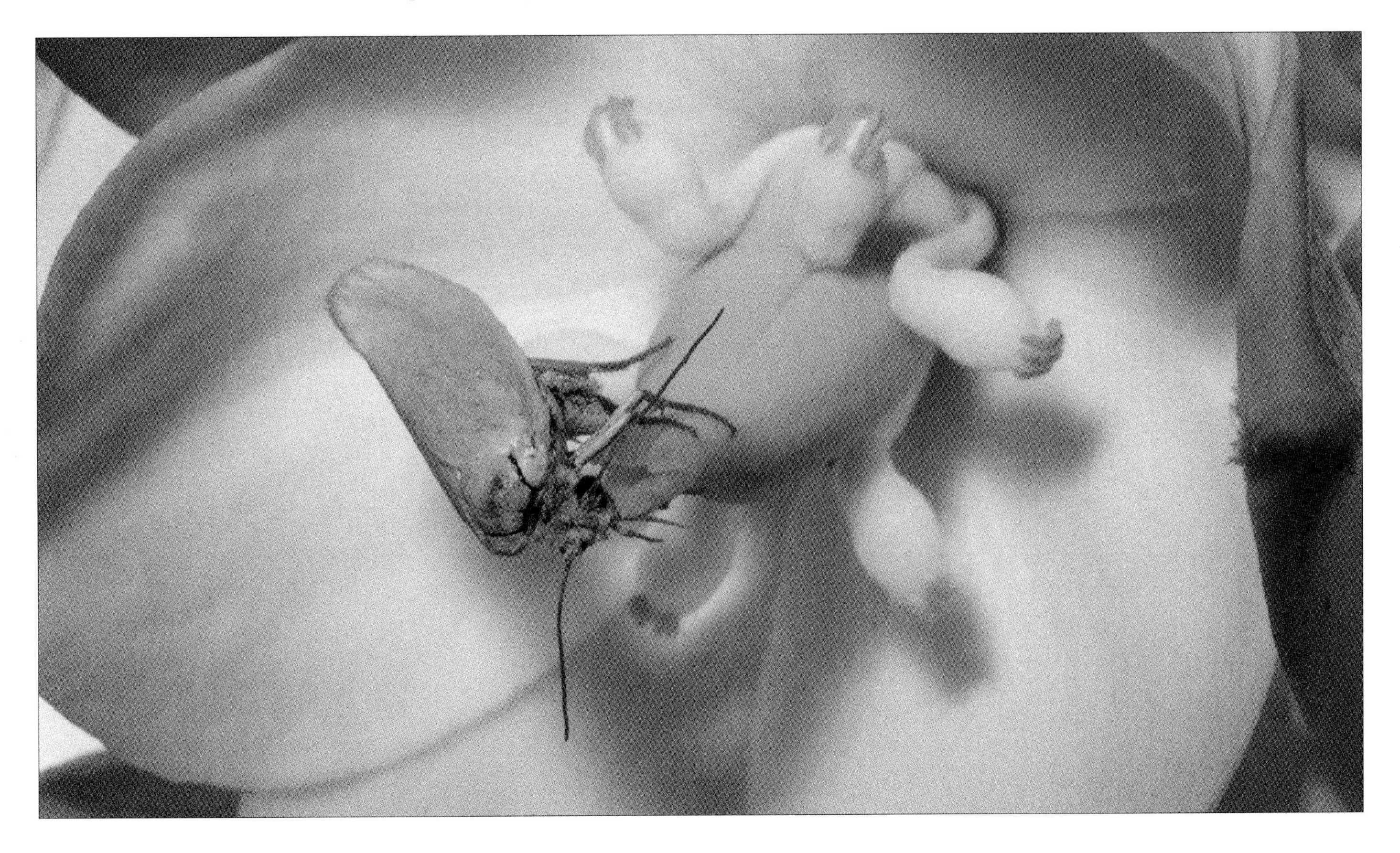

YUCCA MOTH

PHYLUM **Arthropoda**

CLASS **Insecta**

ORDER **Lepidoptera**

FAMILY **Prodoxidae**

GENUS **2: *Tegeticula; Parategeticula;***

SPECIES **More than 100, including *T. yuccasella***

LENGTH
Head and body: usually ⅖ in. (1 cm); wingspan: 1 in. (2.5 cm)

DISTINCTIVE FEATURES
Generally day-flying; drab colors; adapted mouthparts for pollen collection; toughened and slender ovipositor (egg-depositing organ)

DIET
Yucca plant seeds (larvae only)

BREEDING
Age at first breeding: not known; breeding season: not known; number of eggs: 1 or 2; hatching period: not known

LIFE SPAN
Not known

HABITAT
Mainly desert regions, near host plants

DISTRIBUTION
North America

STATUS
Locally common

The female yucca moth lays her eggs in the ovary of a yucca flower using her ovipositor. The larvae feed on some of the yucca seeds, leaving others to start new yucca plants.

the same species and presses the ball of pollen down onto the stigma of the female part of the flower. The moth then lays one or two eggs in the ovary, at the base of the style. She may repeat this six times in a single flower. Then she flies off to repeat the process with other flowers. Some species of yucca moths lay the eggs first and then pollinate the flowers. The pollen fertilizes the ovules, which develop into seeds on which the larvae feed. Because more seeds are produced than the larvae need, some ripen and grow into new yucca plants.

The moth can perform this operation because of two unusual structural features. One is the modification of the mouthparts. The other is the possession by the female of a piercing ovipositor, a specialized organ for depositing eggs, which is highly unusual among moths.

Insects and flowers

Flowering plants and insects have been evolving side by side since the Cretaceous period, about 100 million years ago, when dinosaurs still walked on Earth. The forms, colors, sweet scents and pollen of flowers evolved to attract insects and other pollinating organisms, such as fruit bats, other invertebrates and birds.

Probing into the flowers and flying from one to another, the insects become dusted with pollen, and facilitate the fertilization of one plant by another. This is insect pollination at its simplest level, but precise adaptations of flowers to secure pollination with greater certainty are quite frequent. Adaptations on the part of the insect, as with the transformation of the yucca moth's proboscis, are more rare. Usually it is no more than a simple lengthening of the proboscis to reach into tubular flowers, best seen in some of the hawk moths (discussed elsewhere).

Yuccas are not the only plants that depend on a particular species of insect for their pollination. Many orchids, for example, have flowers that mimic a particular type of insect. The flowers attract male insects, which try to mate with the flowers and so pick up the orchid pollen to carry from one flower to another.

Most associations between yucca moths and plants are symbiotic, in that both parties benefit. However there are other yucca moths that lay their eggs in the fruit rather than the flowers, and do not pollinate their host plant. There are also members of this family of yucca moths that pollinate the plants entirely accidentally in the process of feeding on the flower's nectar.

ZAMBEZI SHARK

THIS SHARK, WHICH IS REMARKABLE for its ability to live in fresh water, has many aliases. In South Africa it is called the Zambezi, shovelnose gray, slipway gray or van Rooyen's shark. In Australia it is called the whaler shark. In Central America it is known as the Lake Nicaragua shark, and in the United States it is the bull, cub or ground shark.

The Zambezi shark measures up to 11½ feet (3.5 m) long and weighs up to 670 pounds (304 kg). It has a broad head with a short, rounded snout and small eyes. The body is heavy with a prominent, triangular first dorsal fin and a small second dorsal fin set well back toward the tail fin, which has a long upper lobe. The pectoral fins are relatively large. The teeth are triangular with serrated edges. The back and flanks are pale to dark gray, and the underside is white.

Widespread in warm oceans, including the Atlantic, Indian and Pacific, the Zambezi shark also frequents inshore waters and enters estuaries and rivers. Individuals have been caught 300 miles (480 km) up the Zambezi. In America, the bull or cub shark is also known to go more than 100 miles (160 km) up rivers. In Central America, Lake Nicaragua, which is 96 miles (154 km) long by 30 miles (48 km) across and 106 feet (32 m) above sea level, has a permanent freshwater population.

Fresh attitude

Wherever it is found, the Zambezi shark lives up to its reputation for attacking large fish (including other sharks), human bathers and even boats. There are several records of canoes being bumped in the Limpopo River in southern Africa, and of one shark biting through the fabric of a canoe so that it eventually sank. There seems to be a connection between shark attacks and the presence of fresh water. Off the Natal coast it has been noted that when floodwaters from rivers drain into the sea, bathers in the brackish water have been attacked. In support of this, there appears to be a greater incidence of attacks on bathers in the freshwater Lake Nicaragua than in coastal waters. Normally a fairly slow-moving shark, the Zambezi shark can put on a sudden burst of speed to attack.

The name of ground shark, by which this species is sometimes known, gives the misleading impression that it spends a significant amount of time resting on the bottom, as some sharks are in the habit of doing. However research in saltwater aquaria, such as that in Durban, South Africa, has shown that unless it is injured or diseased the Zambezi shark does not lie around. If it does, it soon dies from lack of oxygen. Normally the shark swims continuously around the tank, night and day, to maintain the flow of water across its gills.

Catlike vision

Sharks traditionally find their human victims by smell, especially following blood trails in the water. They must use sight as well, and this is emphasized by the accounts of persistent, seemingly deliberate attacks on boats. Although the eyes of the Zambezi shark are small, its sight can be assessed by studies made on related species in the same family. These studies show the eye to be well adapted for vision in bright or dim light. The retina is rich in rods but poor in cones, indicating a high degree of visual sensitivity, poor color vision and poor discrimination of detail. A canoe would therefore be merely a fishlike shape, and one to be attacked. Another feature of the shark's eye is the tapetum lucidum, a layer of

The Zambezi shark, often called the bull shark,* Carcharhinus leucas, *is probably the most dangerous species of tropical shark known to attack humans.

A Zambezi or bull shark, accompanied by remoras, cruising in shallow waters off the Bahamas. This species favors warm oceans and seas.

silvered, mirrorlike plates that reflects light rays back through the retina so that the shark, like members of the cat family on land, can make full use of light of low intensity. As with cats, the pupil is a vertical slit in bright light but open and rounded in subdued light.

Although known for attacking bathers and boats, the real prey of the Zambezi shark is other fish, especially rays and other sharks.

Breeds in fresh water?

Like many sharks, this species gives birth to fully hatched, free-swimming young. Zambezi sharks flock into the St. Lucia estuary in Natal during November and December. It is assumed that they are doing so to give birth in fresh water, for the smallest, youngest sharks are those seen highest upstream, leaping from the water. If this seasonal migration is indeed associated with breeding, then it may explain the shark's heightened hostility in fresh water. Breeding is a time when many animals are at their most aggressive.

Danger from orcas

Tests on several sharks, including the Zambezi shark, have shown that they respond to sounds over a wide frequency and are able to determine the direction of the sounds. The sharks were conditioned to associate sounds with food. Then an interesting experiment was carried out with a tape recording of the sounds made by a group of killer whales 1,800 feet (540 m) away.

When these high-pitched sounds were transmitted into the Shark Research Tank at Durban, none of the sharks in the tank reacted to them except for a large Zambezi shark. This shark seemed to be agitated and swam rapidly around the tank. The killer whale is the most powerful predator in the sea, and this experiment could indicate that the Zambezi shark is one of its habitual victims.

ZAMBEZI SHARK

CLASS **Elasmobranchii**

ORDER **Carcharhiniformes**

FAMILY **Carcharhinidae**

GENUS AND SPECIES ***Carcharhinus leucas***

ALTERNATIVE NAMES
Bull shark; shovelnose gray shark; slipway gray shark; van Rooyen's shark; cub shark; ground shark; whaler shark; Lake Nicaragua shark

WEIGHT
Up to 670 lb. (304 kg)

LENGTH
11½ ft. (3.5 m)

DISTINCTIVE FEATURES
Massive shark; grayish above, white below; broad, blunt snout; small eyes; broad, triangular first dorsal fin

DIET
Bony fish, sharks, rays, mantis shrimps, crabs, squid, sea urchins, sea turtles, carrion

BREEDING
Age at first breeding: not known; breeding season: late spring–early summer; gestation period: 10–11 months; number of young: 1 to 13; breeding interval: not known

LIFE SPAN
16 years

HABITAT
Warm, shallow coastal waters and fresh water, in oceans, seas, bays, rivers and lakes

DISTRIBUTION
Atlantic, from Massachusetts and Morocco south to Brazil and Angola; Indian Ocean and western Pacific, from Kenya to northern Australia; eastern Pacific, from California to Ecuador or Peru

STATUS
Not listed as threatened

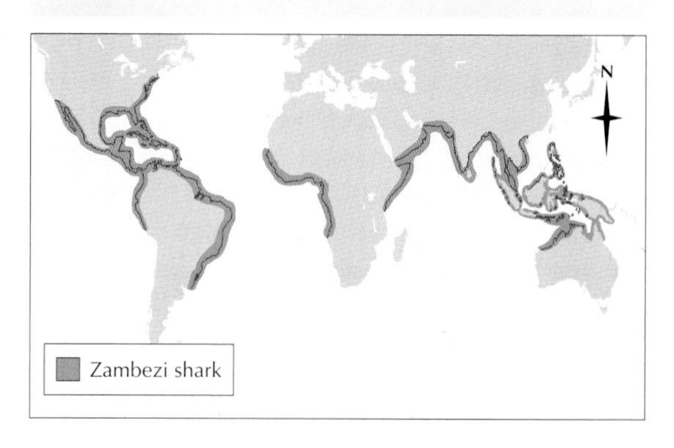

ZEBRA

THEIR BOLDLY STRIPED patterning gives zebras an unmistakable appearance. Generally horselike in appearance, they share features both with wild asses (discussed elsewhere) and horses. Like the asses, zebras have an upright mane and a tufted tail, and have hard wartlike knobs known as chestnuts on the forelegs only. By contrast, horses have chestnuts on the hind legs as well as the forelegs. Both horses and asses exhibit differences from the zebras in the structure of their skull and teeth.

Three species of zebras live in Africa today. The most common and best-known species is the plains zebra, *Equus burchelli*, which has a range extending from southeastern South Africa and northwestern Namibia north as far as southern Somalia and southern Sudan. In this species, the stripes sometimes reach under the belly, and on the flanks they broaden and bend backward toward the rump, forming a Y-shaped saddle pattern. There are two existing subspecies of the plains zebra. Chapman's zebra, *E. b. antiquorum*, occurs in Angola, Namibia and across northern South Africa. Its ground color is yellowish, the belly is unstriped but there are stripes reaching down the legs, usually to below the knees. Between the broad main stripes of the hindquarters and neck there are lighter, smudge gray alternating stripes, known as shadow stripes.

North of the Zambezi River lives the East African subspecies, Grant's zebra, *E. b. boehmi*. Its ground color is white, the stripes continue all the way down to the hooves and there are rarely any shadow stripes. At about 50 inches (127 cm) high and 500–600 pounds (230–270 kg) in weight, Grant's zebra is smaller than Chapman's zebra. It also has a smaller mane, and in the northern districts this feature has disappeared altogether. Maneless zebras occur in southern Sudan and in parts of Uganda and Somalia.

South and southwest of the plains zebra's range lives the mountain zebra, *E. zebra*, which is about the same size as a plains zebra but with a dewlap halfway between the jaw angle and the forelegs. The mountain zebra's stripes always stop short of the white belly. Its ground color is whitish and although the stripes on the flanks bend back to the rump, as in the plains zebra, the vertical bands continue as well, giving a gridiron effect. The southern subspecies, the stockily built, broad-banded Cape mountain zebra, *E. z. zebra*, is almost extinct and survives on only a few

A herd of Chapman's zebras at a Namibian waterhole. Water and food are seasonally scarce in southern and eastern Africa, causing most wild zebras to migrate periodically.

Grevy's zebra (above) is the largest wild horse and is thought by some to be the most primitive member of the horse family.

private properties. The southwestern subspecies, Hartmann's mountain zebra, *E. z. hartmannae,* is still fairly common. It is larger and longer-limbed than the Cape mountain zebra, with narrower stripes and a buff ground color. The third species is Grevy's zebra, *E. grevyi,* from Somalia, eastern Ethiopia and northern Kenya. Grevy's zebra is a very striking, tall animal. The belly is white and unstriped, and a broad dorsal stripe bisects the hindquarters. On the haunches the stripes from the flanks, rump and hind legs seem to bend toward each other and join up.

Grinding jaws

Although they sometimes browse on leaves and twigs, zebras specialize in grazing. They have been observed to eat more than 50 species of grasses, the animals' large, high-crowned teeth, powered by muscular jaws, being resistant to the grasses' fibrous make-up. Zebras are also specialized for running, with muscular upper legs and slender lower legs ending in hoofed feet. These strong limbs can also deliver a mighty kick, making zebras a potentially dangerous target for predators. Even so, along with wildebeest (discussed elsewhere), zebras are a favorite prey of lions.

Stripes of recognition?

In the past, scientists believed zebras' stripes were a form of camouflage that helped to conceal the animals by breaking up their outline. This belief is no longer so widely held. While the patterning may serve some protective function by confusing predators when a herd is moving at speed, biologists are now more inclined in favor of the idea that the stripes play some sort of role in identification and social interaction. Stripe patterns are individually distinctive. Moreover, zebras without stripes are shunned and have little or no breeding success. This last point could be crucial in maintaining stripes over time.

ZEBRAS

CLASS **Mammalia**

ORDER **Perissodactyla**

FAMILY **Equidae**

GENUS AND SPECIES **Plains zebra, *Equus burchelli*; mountain zebra, *E. zebra*; Grevy's zebra, *E. grevyi***

ALTERNATIVE NAMES
Several subspecies also have names

WEIGHT
385–950 lb. (175–430 kg)

LENGTH
Head and body: 75–102 in. (1.9–2.6 m); shoulder height: 42–61 in. (1.06–1.55 m)

DISTINCTIVE FEATURES
Horselike body; short, broad neck; erect mane; tufted tail; black or sometimes brown stripes on white or yellowish ground

DIET
Mainly grasses; also other plant matter

BREEDING
Age at first breeding: 2 years (female), 5–6 years (male); breeding season: all year; number of young: 1; gestation period: about 370 days; breeding interval: 2 or 3 years

LIFE SPAN
Up to 20 years

HABITAT
Plains zebra: savanna. Mountain zebra: mountain grasslands. Grevy's zebra: arid deserts and open grasslands.

DISTRIBUTION
Plains zebra: South Africa north to southern Sudan. Mountain zebra: southern Africa. Grevy's zebra: Ethiopia, Kenya and Somalia.

STATUS
Plains zebra: common. Grevy's zebra and mountain zebra: endangered.

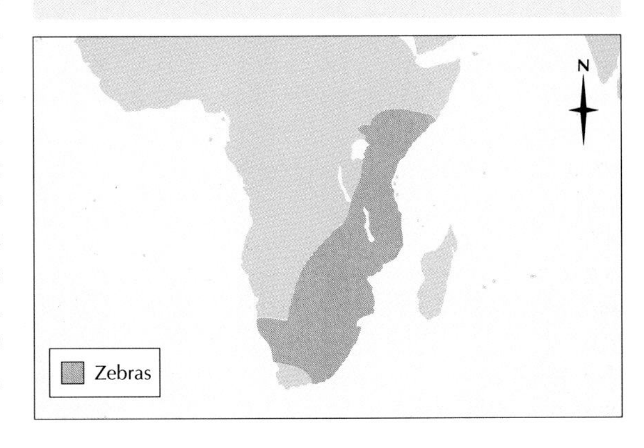

Belligerent stallions

Plains zebras are highly gregarious. Groups of one to six mares and their foals keep together under the leadership of a stallion, which protects them and also wards off other stallions. From time to time, whether through predation, illness or a successful challenge by another stallion, a group's male is replaced. The surplus stallions live singly or in bachelor groups of up to 15 members. Plains zebras are rather tame and do not show as much fear of humans as wildebeest, with which they associate. When they are alarmed, plains zebras utter a barking alarm call, a hoarse *kwa-ha kwa-ha,* ending in a whinny.

Mountain zebras, which are thought to be more aggressive than plains zebras, live in herds of up to six, although sometimes they assemble in large numbers where food is plentiful. They seem to have regular paths over the hills and move along them in single file. The call of these zebras has been described as a low, snuffling whinny, very different from that of the plains zebra. Grevy's zebras associate in family groups and bachelor herds, but the biggest, strongest stallions, which can weigh up to 950 pounds (430 kg), are solitary. Each lone stallion occupies an area of about 1–4 square miles (2.7–10.2 sq km), one of the largest territories of any herbivorous mammal.

Like the other zebra species, Grevy's zebra was hunted for its skin in the past, but the main threat is now loss of habitat to domestic livestock.

A newborn zebra foal has brown stripes and is short-bodied and high-legged, like the foal of a domestic horse. It is born after a gestation period of 370 days, weighs 66–77 pounds (30–35 kg) and stands about 33 inches (84 cm) high. The mares come into season again a few days after foaling, but only a fairly small percentage are fertilized a second time. Usually, a mare has one foal every 2 or 3 years. Mares reach sexual maturity at a little over 1 year old but do not seem to be fertile until they are about 2 years of age. Males leave the herd at 1–3 years old and join the bachelor group. At 5–6 years old, many of them attempt to kidnap young females and, if successful, a new one-stallion herd is formed. The unsuccessful males remain in the bachelor herd or become solitary. Zebras live as long as horses, 15–20 years being typical.

Lost zebras

Two further subspecies of the plains zebra once existed: Burchell's zebra, *E. b. burchelli*, and the quagga, *E. b. quagga*. Both are now extinct. Burchell's zebra was the southernmost subspecies of the plains zebra, occurring in South Africa's Free State and neighboring areas. Burchell's ground color was yellowish, like Chapman's but darker. Also like Chapman's zebra, Burchell's stripes often did not reach under the belly and it had shadow stripes, but its legs were white and unstriped.

The quagga was extremely common in South Africa 150 years ago but was hunted to extinction. This zebra was distinctly striped brown and off-white on the head and neck only. Along the flanks the stripes gradually faded out to a plain brown, which sometimes extended to just behind the shoulders and sometimes reached the haunches. The legs and belly were white. Its barking, high-pitched cry, after which the South African Dutch settlers named it, was seemingly rather like the *kwa-ha* of the plains zebra.

The last quagga died in the Amsterdam Zoo in 1883. However, enthusiasts in South Africa have undertaken a selective breeding program using plains zebras with tendencies toward stripe loss and brown rumps and are slowly creating zebras that resemble quaggas. There is an argument that quaggas are identified only by differences in coat pattern, and that the gene responsible for these differences lies latent in the plains zebra gene pool. Even so, this program will probably produce zebras that merely resemble quaggas, because scientists no longer have either the complete genetic blueprint for the quagga or sufficient information about the animal's true nature.

ZEBRA FINCH

***The zebra finch is a flocking bird. It was formerly placed in the genus* Poephila *but is now classed in* Taeniopygia*, along with the double-barred finch,* T. bichenovii.**

IN THE WILD, THE zebra finch is found only in Australia, but it has become very well known in many countries as a favorite cage bird. It is small and compact and is about 4 inches (10 cm) long. The male is more brightly colored than the female. His back and wings are light brownish gray, the head is dark gray and the sides of the neck and throat are gray finely barred with black, with a black area on the crop. There are distinctive orange cheek patches, separated from the white feathers around the base of the bill by a black stripe. The flanks are light chestnut spotted with white and the underparts are white. The tail is black with broad white bars and the strong bill and legs are bright orange red. The female lacks the male's bright cheek patches and white-spotted chestnut sides as well as the black mark and barrings on the throat. Her underparts have a buffish tinge. A number of color variations or mutations have, however, been bred in cage birds. The zebra is the most common of the Australian finches, being found throughout the continent except for the wet coastal forest areas and on Cape York Peninsula in the north.

Flocking together

Wild zebra finches live in grasslands and savanna woodlands, as well as in scrub, shrublands and orchards and gardens. They are active and hardy little birds that are remarkably unafraid of humans. They live in flocks, which wander about from place to place in search of food. Although they have a variety of calls, zebra finches are not singing birds. They call to each other with a soft, low *tet-tet,* which can be heard all the time the flock is moving about. The identity call is louder and sounds rather like the note from a small toy trumpet. It has also been described as a nasal twang, *tya.* When they become aggressive or chase other birds, zebra finches have a special attack call that resembles the sound of the sudden tearing of a piece of cloth.

Zebra finches never roost in tree branches if they can find nests. Sometimes simple, untidy nests are made for roosting, often with no roof, or else old breeding nests are renovated. Zebra finches feed on or near the ground, chiefly on grass seeds, which are either picked up from the ground or pulled from the stem. During the breeding season, the birds supplement their diet with small insects they take from the ground or catch in flight.

Versatile nester

In common with many other birds of the arid Australian interior, the zebra finch is stimulated to breed by rainfall: no rain, no breeding, usually. It may build its own nest in a thick, prickly bush or in a tree as high as 30 feet (9 m) above the ground. Alternatively, it may use a hole of some kind, an old nest of another bird or even a rabbit burrow. It may even lay its eggs under the large stick nest of a bird of prey. If it builds its own nest, the zebra finch constructs an untidy affair of grasses or small twigs and roots, lined with feathers, plant down and wool. Both male and female build the nest, incubate the eggs and feed the nestlings. From three to seven (usually four or five) pale blue eggs are laid. Occasionally, more than one hen will lay in the same nest, and 23 eggs were recorded in one nest at Marble Bar.

The young birds hatch out in 12–14 days and are fed in the nest for 9–12 days on partially and fully ripe grass seeds regurgitated from the parents' crops. They leave the nest on about the 14th day, but the parents continue to feed them for another 2 weeks, and during this time they usually spend the night in the nest. The fledglings resemble the female in coloration, but the bill is black until the birds are 8–11 weeks of age,

ZEBRA FINCH

CLASS **Aves**

ORDER **Passeriformes**

FAMILY **Ploceidae**

SUBFAMILY **Estrildinae**

GENUS AND SPECIES ***Taeniopygia guttata***

ALTERNATIVE NAMES
Variety of names in captive-bred birds

LENGTH
Head to tail: about 4 in. (10 cm)

DISTINCTIVE FEATURES
Small compact finch; strong, conical bill. Adult male: orange-red bill; bright orange cheek patches; black vertical stripe under eye; dark gray head; brownish gray upperparts; finely barred throat and sides of neck; chestnut flanks with white spots. Adult female: lacks cheek patches, throat barring and chestnut flanks. Juvenile: black bill; otherwise resembles female.

DIET
Mainly seeds; insects in breeding season

BREEDING
Age at first breeding: 6–12 months; breeding season: year-round, according to rainfall, peak usually October–March; number of eggs: usually 4 or 5; incubation period: 12–14 days; fledging period: 13–14 days; breeding interval: 2 to 4 broods per year

LIFE SPAN
Up to 4 years

HABITAT
Grasslands, scrub, shrublands, savanna woodlands, orchards and gardens

DISTRIBUTION
Mainland Australia; absent from wetter coastal districts

STATUS
Very common

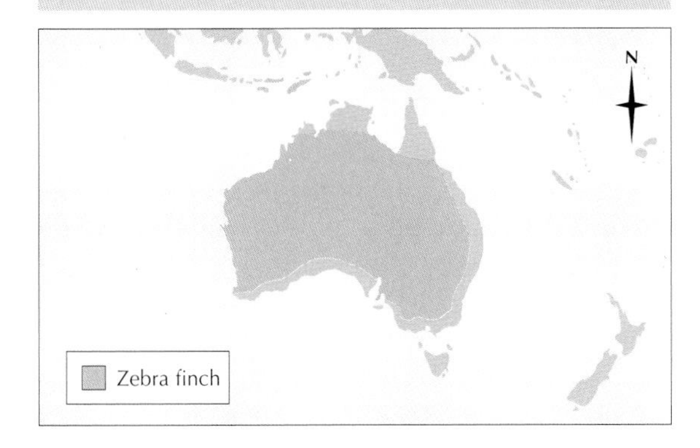

The zebra finch belongs to a group of finches known as grassfinches, most of which occur in Australia.

when it changes to the adult's orange red. Sexual maturity is attained when the young are about 3 months old, but first breeding usually takes place slightly later, usually at about 6–12 months.

Favorite cage bird

The zebra finch is one of the commonest of all cage and aviary birds and one of the easiest to keep and breed in captivity. Although there is now a ban on the export of these birds from Australia, most breeders had already established reliable strains before the ban was imposed, so it is of little disadvantage. A number of color variations have been bred in which the normal gray of the body and wings is replaced by another color, as, for example, in the silver zebra finch and the fawn zebra finch. The white zebra finch is white except for the bill and legs, and the sexes look almost identical.

A number of these hardy and gregarious birds can be kept together in an outdoor aviary year-round, so long as dry, draft-proof sleeping quarters are provided, with perhaps a little heat in the winter months. Breeding birds, however, do best in an indoor aviary. The birds thrive on millet and canary seeds with some fresh green food. They also need water, grit and some form of shell-forming substance.

The zebra finch is the easiest and most prolific breeder in captivity of any of the finches and, given the opportunity, would breed nonstop throughout the year. Too many clutches, however, weaken the hen, which eventually produces less healthy chicks.

ZEBRA FISH

The zebra fish is one of the most popular aquarium fish in the United States and can be kept with other nonaggressive species.

There are several fish with a common name that includes the word "zebra." Among these is a small freshwater fish of India, Pakistan, Bangladesh and Nepal. Up to about 2½ inches (6 cm) long, it is called the zebra fish, or zebra danio, and is a member of the large carp family (Cyprinidae). In the wild it occurs in slow-moving to stagnant water, including streams, canals, ditches, ponds and especially in rice fields. The zebra fish is also an extremely popular fish with aquarists, and occurrences of the species in the wild in Florida and Colombia are probably due to escapes from breeding facilities for the aquarium trade.

The zebra fish is slim, with the body only slightly compressed. The dorsal fin and the anal fins are fairly large, as is the tail fin, while the pelvic and pectoral fins are small. There are two pairs of barbels on the chin. The male's back is brownish olive, the belly is yellowish white and the flanks are a gleaming blue with four gold longitudinal stripes extending from the gill cover to the base of the tail. The gill covers are blue with golden blotches and transverse bars. The dorsal fin is blue, yellow olive at its base with a white tip, and the anal fins are barred with blue and gold, as is the tail fin. The effect of the stripes is to make the zebra fish appear even more streamlined than it is and to give an impression of movement even when it is stationary.

Zebra fish swim among water plants or in schools, and it is when a group is all aligned, swimming in formation, evenly spaced and all traveling in the same direction that the fish most catch the eye. Almost certainly their attraction owes much to the repetition of their stripes, a phenomenon termed "beauty in repetition" by the scientist Dr. Dilwyn John in 1947.

Female less colorful

There is little difference between the sexes except that the female zebra fish is larger and, especially just before spawning, more plump than the male. She is also a little more sober in her coloration, her stripes being more silver and yellow than the gold stripes of the male. As part of the courtship behavior, the male leads the female in among the water plants and the two take up position side by side. The female then sheds her ova (eggs) and the male sheds his milt over them to fertilize them. The eggs sink to the bottom and the adults take no care of them.

Thwarting the egg eaters

There is one aspect of zebra-fish spawning behavior against which aquarium keepers need to take special precautions. Zebra fish are carnivorous, feeding on any small animals they can swallow, usually worms, small crustaceans and insect larvae (zebra fish can be used as a biological agent for mosquito control). However, zebra fish also eat their own eggs, tending to snap them up as they slowly sink after fertilization. The first precaution for the aquarist is therefore to provide a breeding aquarium with water shallow enough that the fish have no chance to catch the eggs before they sink into the safety of the gaps between the gravel on the bottom. Also, the correct size of pebbles must be used or the adults may become trapped between them. Marbles offer one solution to the problem. Another method is to use a trap, such as a false tank bottom of metal or plastic mesh, which allows the eggs to fall to the real bottom of the aquarium, beyond the adults' reach.

ZEBRA FISH

CLASS **Osteichthyes**

ORDER **Cypriniformes**

FAMILY **Cyprinidae**

GENUS AND SPECIES ***Danio rerio***

ALTERNATIVE NAME
Zebra danio

LENGTH
2½ in. (6 cm)

DISTINCTIVE FEATURES
Body very slim, only slightly compressed; two pairs of barbels. Adult male: brownish olive back, yellowish white belly, gleaming blue flanks with 4 gold bands (blue color contained within area bounded by upper and lower band); anal and tail fins banded blue and gold, blue dorsal fin with white tip and yellow-olive base; golden red iris. Adult female: larger and more robust than male; coloration more sober, banding silvery to yellowish.

DIET
Worms, small crustaceans, insect larvae; fish eggs

BREEDING
Age at first breeding: 1 year; breeding season: regularly throughout year; number of eggs: about 200; hatching period: 2 days

LIFE SPAN
3 years

HABITAT
Streams, canals, ditches, ponds, especially rice fields

DISTRIBUTION
Asia: Pakistan, India, Bangladesh and Nepal; also Florida and Colombia (probably escapes or releases)

STATUS
Not threatened

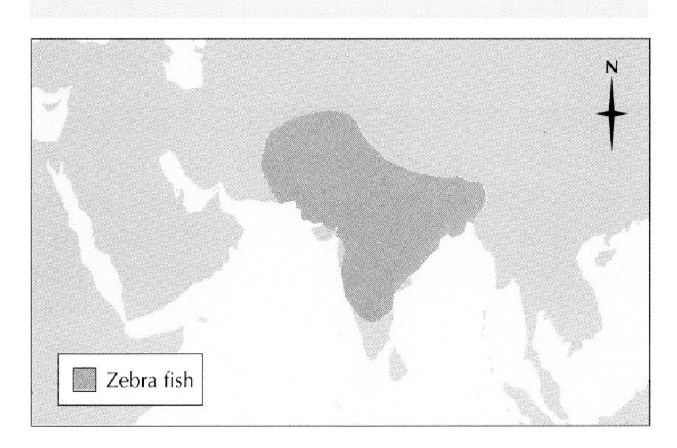

Zebra fish are common in Southeast Asia. They favor slow-moving or still waters, especially in rice fields.

Each female lays about 200 eggs, which hatch in 2 days. The larvae are at first fairly helpless and inactive, but after a further 2 days they can swim and start to feed on microscopic plankton animals. Zebra fish begin to breed at 1 year old. At 2 years they are old-aged, and a zebra fish of 3 or more years is an extreme rarity.

A question of stripes

Originally reserved for the striped wild horses of Africa, the term "zebra" now occurs in the names of numerous animals to indicate stripiness. In the world of fish there is the zebra shark, *Stegostoma fasciatum*, of the Indian Ocean, which has black or brown bars on the body. In the extreme south of South America is the zebra salmon, *Aplochiton zebra*, while a marine fish of the Indian and Pacific Oceans, the Jarbua terapon, *Terapon jarbua*, is sometimes called the zebra or tiger fish.

Among the aquarium species, besides the zebra fish, there is the striped, or zebra, barb, *Puntius eugrammus*, of Peninsular Malaysia and the East Indies. There is also the plains killifish, *Fundulus zebrinus*, of North America, which is also called the zebra killie, and the zebra cichlid, *Cichlasoma nigrofasciatum*, also known as the convict cichlid.

ZEBRA MOUSE

The four-striped grass mouse,* Rhabdomys pumilio, *is named for the black stripes running along its back.

THE ZEBRA MOUSE IS found only in Africa, where it is widespread throughout most of the continent. Also known as the striped grass mouse, the term zebra mouse is actually used to describe a number of species belonging to the genus *Lemniscomys*. Scientists recognize at least 10 separate species, but much of the information about these small rodents comes from four species (*Lemniscomys striatus, L. barbarus, L. griselda* and *L. macculus*), which differ mainly in the coloring and marking of their fur. *L. griselda* is plainly colored, *L. striatus* and *L. macculus* are both spotted, and *L. barbarus* is characterized by small stripes. In all species, the basic color is usually light brown.

Zebra mice have long and slender bodies, which vary in size according to species. Both *L. griselda* and *L. striatus* have an average head and body length of about 4½–5 inches (11.3–12.5 cm), a tail of the same length, and weigh about 1¾ ounces (50 g). Slightly smaller, *L. barbarus* and *L. macculus* have a head and body length of about 4 inches, a similar tail length and weigh only 1 ounce (30 g). The head is long and blunt-nosed, with no obviously defined neck area. The eyes are large, black and elliptical, and the ears are small and petal-shaped. The belly may vary from white or buffy white to pale gray, and the feet are light brown or white. The animals are covered by thin, coarse fur, which becomes transparent along the tail. In some species, the tail is bicolored, with a paler underside. Powerful hind legs make these mice excellent jumpers, and they move with a rabbitlike hopping motion.

Differing habitats

Zebra mice live in a variety of habitats, ranging from riverbanks and reedbeds to savanna and semidesert. In Ghana, *L. striatus* is associated with the grasslands found in forest clearings. The range expansion of some zebra mouse species is likely to have been favored by the migration of people over the past few centuries. Indeed, pastoralists probably created corridors of grassland by lighting fires, thus benefiting the spread of these rodents. Today, species from the genus *Lemniscomys* range from north of the Sahara to South Africa and from West Africa through to southeast Kenya and Tanzania.

While some species, such as *L. barbarus*, are found over wide areas of Africa, others have a more limited distribution. For example, *L. mittendorfi* is found only in Cameroon, where it is considered endangered. *L. griselda* is also rare throughout much of its east African range, although it is locally abundant in some woodlands, and common over southern Africa. Some species, such as *L. striatus*, also thrive in mountainous areas, at elevations of up to 11,550 feet (3,500 m). Due to the wide variety of habitats used by zebra mice, it is often only possible to identify the ecological niches occupied by the different forms in regions where more than one species occur. For example, where *L. striatus* and *L. barbarus* coexist, the former tends to favor moist drainage lines, while the latter prefers grasslands with efficient run-off.

Secretive lives

Zebra mice prefer to shelter underground, favoring burrows or abandoned termite nests. Although timid, they are active mainly during the day, usually resting from the sun during the hottest part of the day. Predators include snakes, small cats, mongooses and birds of prey, such as hawks. They are seldom present in owl pellets, a reflection of their daytime activity patterns. When they are disturbed, these mice reportedly jump straight into the air before scuttling away hurriedly. Foraging for food occurs during dawn and dusk. Zebra mice are omnivorous, feeding on plant material such as leaves, roots, fruit and

ZEBRA MICE

CLASS **Mammalia**

ORDER **Rodentia**

FAMILY **Muridae**

GENUS AND SPECIES ***Lemniscomys barbarus; L. striatus; L. macculus; L. griselda; L. bellieri; L. hoogstraali; L. linulus; L. mittendorfi; L. rosalia; L. roseveari***

ALTERNATIVE NAME
Striped mouse

WEIGHT
1–1¾ oz. (30–50 g), depending on species

LENGTH
Head and body: 4–5 in. (10–12.5 cm); tail: 4–5 in. (10–12.5 cm)

DISTINCTIVE FEATURES
Slender body; long tail; short, coarse fur; striped, spotted or plain fur on back

DIET
Omnivorous: leaves, roots, fruit, seeds; insects and other invertebrates

BREEDING
Gestation period: about 3–4 weeks; number of young: about 5 per litter

LIFE SPAN
Up to 3 years in captivity

HABITAT
Very varied, including savanna, semidesert, reedbeds, forests, riverbanks

DISTRIBUTION
Most of Africa; species vary with areas

STATUS
Most species fairly common

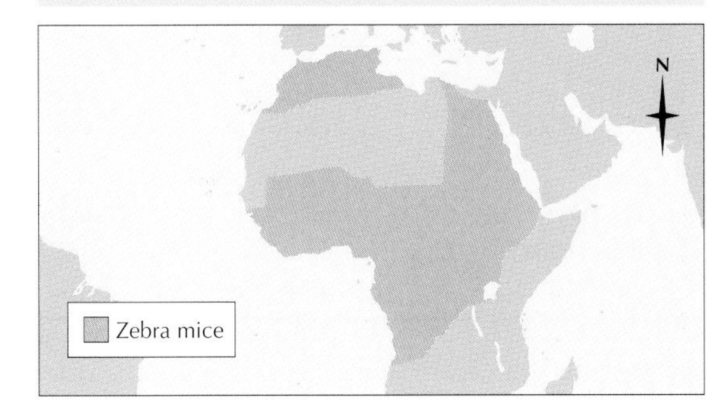

seeds, as well as insects and other invertebrates. They have well-defined runs leading through the grass and small piles of grass stems provide evidence of feeding. In some areas these rodents are viewed as pests, because they feed on crop plants such as bananas, cassava and potatoes.

Zebra mice range across much of Africa. R. pumilio *is found in the cooler grasslands of South Africa and in the mountains of east Africa.*

Prolific breeders

Some species of zebra mice construct a dome-shaped nest from grass, in which the offspring are born and cared for. While some species appear to breed year-round, others have a more defined breeding season, mating to coincide with the rainy season. Estrus (receptivity to mating in the female) lasts about 5 days, and gestation lasts 21–28 days. Although litters usually comprise two to five pups, litters of more than 10 individuals have been recorded. When conditions are favorable, zebra mice are prolific breeders. A breeding pair may produce 21 young in four litters in just 4 months. Newborn offspring weigh about ⅒ ounce (3 g) and are born with a thin coat on which the stripes are already visible. The eyes open after about 1 week and the mice reach adult size after 5 months. They become sexually mature after just a few months, although females do not breed until 1 year old. Zebra mice are kept in captivity in many zoos, where they live from 2–3 years. Recently, these rodents have also become popular as house pets. Little is understood about the social behavior of these species in the wild. Most reports indicate they are not gregarious animals, although they do sometimes live in close proximity to each other.

A complex group

The genus *Lemniscomys* is a complex group. While several species of zebra mouse are recognized, many subspecies have also been described. These races may differ from each other very subtly in their morphology, or even in their chromosomal arrangement. The problem of species identification in this genus has been highlighted by a recent study, which showed that based on skull and teeth features, populations currently assigned to *L. barbarus* actually comprise two species. Future genetic research would help to resolve the taxonomy of this widespread mammal.

ZEBU

Some zebu have impressively sized horns, up to several feet long. Others have very short horns or none at all.

THE ZEBU OR BRAHMAN cattle originated in southern Asia, possibly in India, but their wild ancestor is not known. The zebu, *Bos indicus*, has a prominent hump on the shoulders, which is an enlarged muscle rather than a store of fat, as many people believe. Its coat is usually gray, although it may be white or black; its legs are slim and its horns are more upright than those of the aurochs, the wild ancestor of European cattle (discussed elsewhere). It has a marked dewlap and drooping ears. Some biologists claim it was domesticated from the wild gaur, *B. gaurus*, or the banteng, *B. banteng*, of Southeast Asia. However, the zebu differs from these in its long, slender face and other features.

Domestication of the zebu may have occurred earlier than that of western cattle. The first record was made in 4500 B.C.E., but on wax seals from West Pakistan, dated 2500 B.C.E., there are representations of both zebu and western cattle. The two types of cattle readily hybridize, and offspring from such crosses have produced some of the breeds found in Africa, where the zebu is better adapted to the hot climate.

Producing hybrids

Much research has been directed toward making more use of such crosses. The western cattle are better for milk production, but this and meat production decline in animals taken from temperate to tropical countries. Zebu are usually better for meat production. In addition, the zebu has larger ears than western cattle, and these and its dewlap offer a large surface for loss of body heat by radiation.

In other words, zebu can, because of their build, keep cooler than western cattle, which, if taken to the Tropics, spend less time grazing and more time lying in the shade, leading to malnutrition. This is enhanced by their woolly coat, which also inhibits heat loss. Thus, they lose more water through their breathing, which, among other things, lowers their milk yield. In addition, zebu also sweat freely through pores in their skin, which adds to their extremely impressive heat tolerance.

Much modern biological research is aimed at producing hybrids that combine the most advantageous characteristics from the two species for giving maximum milk or meat production, or both. However, hybridizing the two species has been carried out repeatedly in different places in the course of history. Sanga cattle, widespread in Africa, probably originated in Egypt from crossing Hamitic longhorn cattle with zebu.

Gentle and docile

There is a distinct difference in temperament between the two species of cattle. The zebu is more tractable and docile, yet it is more lively than western cattle, and grunts rather than moos. When it is walking, it swings its hind legs in a straight line, in the manner of a horse, instead of using the sideways movement of the dairy cow.

The zebu has endurance and speed, and zebu are recorded as carrying a soldier on their back for 16 hours a day at a speed of 6 miles per

ZEBU

CLASS **Mammalia**

ORDER **Artiodactyla**

FAMILY **Bovidae**

GENUS AND SPECIES ***Bos indicus***

ALTERNATIVE NAME
Brahman cattle

WEIGHT
Male: 1,600–2,200 lb. (726–1,000 kg); female: 1,000–1,400 lb. (453–635 kg)

LENGTH
Shoulder height: up to 6 ft. (1.8 m)

DISTINCTIVE FEATURES
Hump above shoulders due to enlarged muscle; prominent dewlap; drooping ears; upright horns; gray, white or black coat; slender legs; extremely tolerant of heat

DIET
Grasses and hay

BREEDING
Age at first breeding: about 2 years; breeding season: induced by long day-length; number of young: 1; gestation period: about 290 days; breeding interval: 1 year

LIFE SPAN
About 15 years

HABITAT
Kept widely in hot countries

DISTRIBUTION
Central and eastern Africa; Indian subcontinent; smaller numbers and hybrids in U.S. and Australia

STATUS
No status (domestic breed)

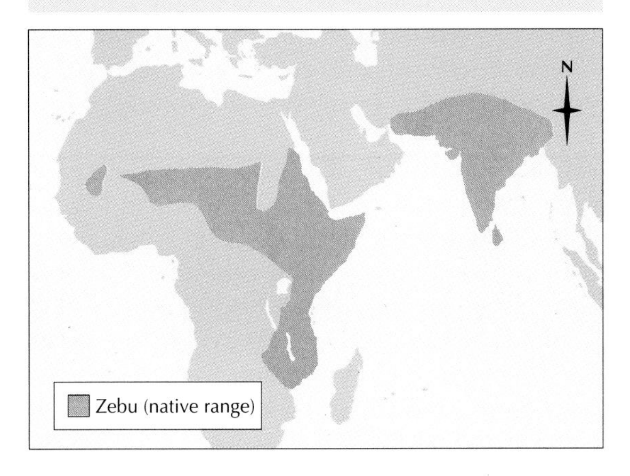

Zebu surrounded by cattle egrets, Pantanal, Mato Grosso do Sul, Brazil. These cattle are of extreme economic importance in many countries.

hour (9.6 km/h). Zebus were frequently used for drawing vehicles, from state carriages to heavy farm wagons or plows, and are still used today for transporting heavy loads.

Zebu are said to be capable of clearing a five-barred gate with ease. One observer tells of a calf that would leap over an iron fence to reach water, and when it had drunk its fill, would leap back again.

Tick- and heat-resistant

The zebu has been taken across the tropical and subtropical regions of the Old World, from China to Africa. It was taken to the United States about 150 years ago and also to Australia, primarily to the state of Queensland in the northeast.

The zebu's popularity and that of hybrids with western cattle in the United States rests not only on its ability to stand up to heat but also on its immunity to ticks that carry Texas fever. The hybrids share this immunity, at the same time producing better meat than the pure zebu.

The size of zebu ranges from that of a small donkey to animals larger than any western cattle. Their horns vary considerably in form, from hornless and small-horned, with horns shorter than the ears, to those of the Ankole cattle of Uganda, which are enormous, stout, swept-back horns that grow up to 5 feet (1.5 m) long.

ZORILLA

Three white patches emblazon the head of the zorilla, part of its dramatic warning to other animals not to interfere with it.

THE ZORILLA OR STRIPED POLECAT of Africa closely resembles the American striped skunks, genus *Mephitis*, although it is only distantly related to them. A small, slender member of the weasel family, it has a head and body length of 11–15 inches (27.5–37.5 cm) and a tail 8–12 inches (20–30 cm) long. Its long fur is strikingly marked in black and white. The body is black with broad white stripes down the back, the face is black with three large white spots, and the bushy tail is mainly white. Like the skunks, the zorilla reacts to disturbance by ejecting a foul-smelling fluid from its anal glands.

One of the most common mammals of Africa, the zorilla ranges from Senegal, northern Nigeria, the Sudan and Ethiopia to southern Africa. There is only one species, though many local races have been described. Two slightly smaller relatives of the zorilla in Africa resemble it in having black-and-white markings. They are the Libyan striped weasel, *Poecilictis libyca*, from North Africa and the white-naped weasel, *Poecilogale albinucha*, from central South Africa. They are known as snake-weasels, but whether this is because they kill snakes or because their low, flat head resembles the head of a puff adder has never been established.

Potent defense mechanism

The zorilla is found in a variety of habitats but it seems to favor dry areas. It is usually solitary, except in the breeding season. During the day it shelters in a rock crevice or a burrow dug with its long, strong claws. It sometimes uses the burrows of other animals or even shelters under farm buildings or outhouses. The zorilla hunts during the night, usually on the ground, but if necessary it can climb or swim after its prey. It trots slowly along on its short legs with its back slightly hunched, waving its long, bushy tail.

At the first sight of a potentially dangerous intruder, the zorilla bristles the hair on its body and, turning its back toward the intruder, stiffly raises its tail. If it is attacked, it ejects fluid from glands that open each side of its anus. Apparently, it does so with impressive accuracy, usually into the face of the attacker. The fluid is potent and can cause temporary blindness. If this is ineffective, however, the zorilla may feign death to escape being mauled or killed. There seems to be a difference of opinion about the fluid. Some naturalists argue it is less pungent than that of the American skunks, while others claim its odor is worse than that of any other animal. It is possible that the odor may vary with the age of the individual or perhaps even with the time of year. Peoples of the Sudan are, however, in no doubt, and have named the zorilla the "father of stinks."

Expert ratter

The zorilla's diet consists mainly of mice and other rodents, small reptiles, birds and large insects. It follows burrowing mole rats underground to catch them. It also takes eggs and snakes from time to time. Although it may occasionally kill poultry, the zorilla does more good than harm around farms in keeping down rodents. Its food is bolted down in lumps rather than chewed, and, as with cats, does not hold each morsel with the paws before eating it. The zorilla kills more than it needs for immediate consumption. A habit it shares with other small members of the weasel family is that of hiding surplus food; it piles the carcasses in a neat heap.

ZORILLA

CLASS **Mammalia**

ORDER **Carnivora**

FAMILY **Mustelidae**

GENUS AND SPECIES ***Ictonyx striatus***

ALTERNATIVE NAMES
Zorille; striped polecat

WEIGHT
1–3 lb. (420–1,400 g)

LENGTH
Head and body: 11–15 in. (27.5–37.5 cm); tail: 8–12 in. (20–30 cm); male usually larger than female

DISTINCTIVE FEATURES
Long, thick, glossy black coat with broad white stripes down back; black face with three large, white spots; tail mostly white

DIET
Rodents; birds and their eggs; amphibians and reptiles, such as snakes; insects

BREEDING
Age at first breeding: 10 months (female), 22 months (male); breeding season: spring and summer; gestation period: 36 days; number of young: 1 to 3; breeding interval: 1 year, second litter may follow if first litter lost

LIFE SPAN
Up to 13⅓ years in captivity

HABITAT
Wide range of savanna types; favors open grassland in high, dry terrain; avoids dense forest

DISTRIBUTION
Most of Africa, from Senegal and Ethiopia south to South Africa

STATUS
Common

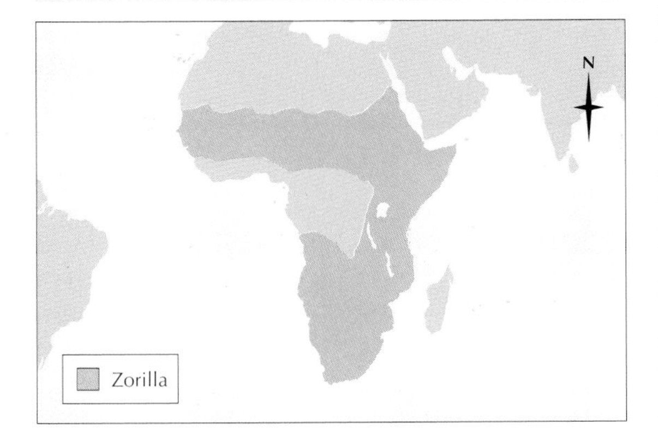

***Along with the coat pattern, the bushy tail marks this species apart from other two-toned mustelids of Africa, such as the striped weasels* Poecilogale *and* Poecilictis.**

Growth of the young

Zorillas breed from early spring to late summer. About 5 weeks after mating, a female gives birth in a burrow, usually to two or three young. The offspring, blind at birth, each weigh just over ½ ounce (14 g). Their eyes open at 40 days, by which time they have started to take solid food. Weaned at 18 weeks, they are fully grown a couple of weeks later.

If it is taken young, a zorilla can make a friendly and gentle pet, seldom emitting its fluid except when frightened. This is a common experience with other mustelids, such as the European polecat, that are also able to give out a noxious fluid. Zorillas have in the past occasionally been kept in zoos; one captive is reported to have lived for 13 years and 4 months.

An intimidating opponent

Although the zorilla may be attacked by dogs or larger predators, most animals, warned by its striking coloring, give it a wide berth to avoid being sprayed by its glandular fluid.

Ivan T. Sanderson, in his *Living Mammals of the World*, quotes a famous game warden for the story of how a zorilla drove nine adult lions away from a freshly killed zebra. The zorilla had taken possession of the carcass and for several hours sniffed around it, occasionally nibbling at it and even taking a nap with its back to the zebra. Whenever a lion attempted to approach the carcass, the zorilla raised its tail in warning and the big cat retired, apparently wary of the stream of amber-colored liquid that might follow.

Index

Page numbers in *italics* refer to picture captions.
Index entries in **bold** refer to guidepost or biome and habitat articles.

Page numbers in *italics* refer to picture captions. Index entries in **bold** refer to guidepost or biome and habitat articles.

Page numbers in *italics* refer to picture captions. Index entries in **bold** refer to guidepost or biome and habitat articles.